MORE PHYSICS WITH MATLAB

Note from the Publisher

**More Physics with MATLAB
(With Companion Media Pack)**
ISBN: 978-981-4623-93-3 (hardcover)
ISBN: 978-981-4623-94-0 (paperback)

The Companion Media Pack is available online at http://www.worldscientific.com/worldscibooks/10.1142/9330#t=suppl

1. Go to http://www.worldscientific.com/r/9330-supp
2. Register / login/
3. You should be re-directed to
   /http://www.worldscientific.com/worldscibooks/10.1142/9330
4. Click on the Supplementary tab to download the Media Pack

MORE PHYSICS WITH MATLAB

Dan Green
Fermi National Accelerator Laboratory, USA

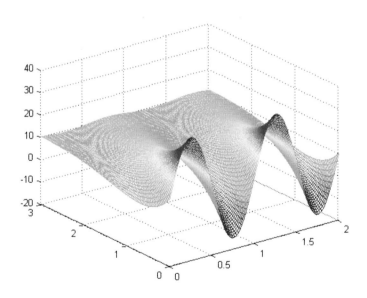

World Scientific

NEW JERSEY · LONDON · SINGAPORE · BEIJING · SHANGHAI · HONG KONG · TAIPEI · CHENNAI

Published by

World Scientific Publishing Co. Pte. Ltd.
5 Toh Tuck Link, Singapore 596224
USA office: 27 Warren Street, Suite 401-402, Hackensack, NJ 07601
UK office: 57 Shelton Street, Covent Garden, London WC2H 9HE

British Library Cataloguing-in-Publication Data
A catalogue record for this book is available from the British Library.

MORE PHYSICS WITH MATLAB
(with Companion Media Pack)

Copyright © 2015 by World Scientific Publishing Co. Pte. Ltd.

All rights reserved. This book, or parts thereof, may not be reproduced in any form or by any means, electronic or mechanical, including photocopying, recording or any information storage and retrieval system now known or to be invented, without written permission from the publisher.

For photocopying of material in this volume, please pay a copying fee through the Copyright Clearance Center, Inc., 222 Rosewood Drive, Danvers, MA 01923, USA. In this case permission to photocopy is not required from the publisher.

ISBN 978-981-4623-93-3
ISBN 978-981-4623-94-0 (pbk)

Printed in Singapore by Mainland Press Pte Ltd.

"It should be possible to explain the laws of physics to a barmaid."
— **Albert Einstein**

"Not only is the Universe stranger than we think, it is stranger than we can think."
— **Werner Heisenberg**

Preface

"Computers are like Old Testament gods; lots of rules and no mercy."
— **Joseph Campbell**

"Part of the inhumanity of the computer is that, once it is competently programmed and working smoothly, it is completely honest."
— **Isaac Asimov**

There are only a very few analytically solvable problems in physics. They are extremely useful because the equations for the solutions can be plotted and the parameters defining the solutions can be varied in order to explore the dependence of the solutions on the variables of the problem, although that exercise may become tedious. In that way the student can build up an intuition about the Kepler problem, for example.

However, this can only be done in a few cases and even then the effort needed is large. For the others, numerical methods are needed and the computation becomes somewhat cumbersome. As a result, it is more difficult to vary the parameters of the problem numerically rather than symbolically and develop an intuition about the dependence of the solution on those parameters. In particular the time development of the system is often obscure and "movies" can be a welcome tool in improving physical intuition.

Because of these issues, the advent of powerful personal computing has considerably reduced the difficulties. Indeed, the aim of this book is to use the ensemble of symbolic and numeric tools available in the MATLAB suite of programs to illustrate representative numerical solutions to more than one hundred problems spanning several physics topics. The student typically works through the demonstration and alters the inputs through a menu driven script. In that way the user driven menu allows for parametric variation.

The more advanced student can edit the script or, indeed, write a new script covering a wholly new problem. It is crucial that the student read the script which is supplied with the text. There are many "comment" lines in the script which more fully explain what the procedures are in each specific case.

MATLAB is a good vehicle for the computational tasks. It has a compiler, editor and debugger which are very useful and user friendly. The HELP utility is very extensive. The MATLAB language is similar to a modern C++ language and it is a vector/matrix language which makes coding simpler than older languages such as FORTRAN.

MATLAB contains many special functions. Matrices and linear algebra are covered well. Curve fitting, polynomials and fast Fourier transforms are supplied. Numerical integration packages are available. Differential equations, symbolic, ordinary and partial, as well as numerical solutions are available for both initial value and boundary value versions.

As an additional package, MATLAB has symbolic mathematics. Within that package, calculus, linear algebra, algebraic equations and differential equations are covered. It is easy to combine a symbolic treatment of a problem with a numerical display of the solution when that is desirable. In this way converting from symbols to numbers is easily achieved.

Finally, and very importantly, MATLAB has an extensive suite of display packages. One can make bar, pie, histogram and simple data plots. There are several contour and surface plots which are possible. There are two- and three-dimensional plots of all types available. The time evolution of solutions can be made into "movies" that illustrate the changes in speed during an evolving process. These extensive visualization tools are crucial in that the student can plot and then vary and re-plot.

There are many topics which are explored in the text. But, it is not what topic is specifically explored that is important but how it is examined. A useful methodology that one can apply to any specific problem is, in outline:

1. First look at the dimensions of an equation to see if they are correct.

For example, the Schrödinger equation has units of energy

$$[-h^2(\partial^2/\partial x^2)/2m + V = ih\partial/\partial t]\psi.$$

Since the Planck constant has dimensions of position times momentum or energy times time, it is clear that all the terms in the equation have the dimensions of energy.

2. Then see what are the characteristic lengths and times implied by the equation.

For the previous equation there is characteristic length times energy, $[L] = hc\sqrt{1/mc^2}$. For electrons, that length is 8 eV ∗ A. The second term has dimensions eV. The third term has a time scale of $0.66 \,\text{eV} \times 10^{-15}$ sec. Therefore for problems with scales of V of a few eV, it is expected that distances for electrons will be a few Angstroms and times will be a few 0.001 psec.

3. See if there are physically reasonable limits implicit in the equation.

For example, if the section on "Rocket Drag" the equation of motion is:

$$d^2y/dt^2 = -v_o(dM/dt)/M - g - k\rho v^2/M$$
$$\rho = \rho_o e^{-\beta y}$$

In the limit where k and g are zero, the simple rocket equation is recovered as a known limit. This can be checked by choosing the parameters of the problem appropriately.

4. Try to solve the equation symbolically, if not use numerical techniques.

5. For either type of solution look for limits which simplify the results and are physically reasonable.

For example, the analytic solution of the Kepler problem for an orbit with eccentricity e is:

$$r = r_c(1+e)/(1 + e\cos\theta)$$

Clearly, a circular orbit has a constant radius, so that such an orbit must have e of zero. At the other limit, if the radius is to diverge, then the cosine of the angle must be $1/e$.

Therefore unbound orbits must have $e > 1$ and the cosine of the angle of the orbit at large r, the asymptote, is $1/e$.
6. Finally, vary parameters of the problem in order to establish an "intuition" about the dependencies of the result on the parameters.

For example, the "Spring Pendulum" is governed by the equations:

$$d^2r/dt^2 = g\cos\theta + (k/m)(Lo - r) + r(d\theta/dt)^2$$
$$d^2\theta/dt^2 = -g\sin\theta/r - 2(dr/dt)(d\theta/dt)/r$$

These are solved numerically. Possible checks are to set g to zero when a simple spring should result. Another limit would be to set k to zero, when a simple pendulum should result. Both limits can be checked by choosing the parameters, including the initial condition, of the problem.

The aim of using these tools is to create intuition, not to solve a specific problem or to complete a specific number crunching exercise. Indeed, the aim of the text is not to teach the basic physics but to give the user a sense of how the solutions of a given physics problem depend on the parameters of that problem. If the reader has questions, a dip into Wikipedia and/or Google should be very helpful. Details are also available in the "comments" lines within the specific script.

The script for these demonstrations is made available on storage media and examples for classical mechanics appear in the Appendix. The student is strongly advised to read the script before executing the script. In that way the script is not a "black box" but is part of the learning process. Using that material the student can write her own additions and explorations with the scripts supplied used as jumping off points. In this way, a path is made available to extend well beyond the specific demonstrations provided in the book itself, making the search for further possible insights open ended.

Contents

Preface vii

1. Mathematics — 1

    1.1 Chaos 1
    1.2 Malthus 3
    1.3 Missile Tracking 4
    1.4 Monte Carlo Analytic 5
    1.5 Monte Carlo Numeric 9
    1.6 Monte Carlo Muon Beam 11
    1.7 Fourier and Laplace Transforms 14

2. Classical Mechanics — 18

    2.1 Angular Momentum 18
    2.2 Classical Scattering 19
    2.3 Impact Parameter and DCA 20
    2.4 Foucault Pendulum 23
    2.5 Spring Pendulum 26
    2.6 Spherical Pendulum 28
    2.7 Driven Pendulum 30
    2.8 Random Walk 32
    2.9 Rotating Hoop 33
    2.10 Tides 36
    2.11 Action 38
    2.12 Rocket Drag 40
    2.13 Two Stage Rocket 43
    2.14 Table Top 46
    2.15 Kepler Eccentric 48

	2.16	Non-Central Force	48
	2.17	Euler Angles	51
	2.18	Forceless Top	53
	2.19	Top Motion	53

3. Electromagnetism   59

	3.1	Electron e/m	59
	3.2	Potential and Fields	61
	3.3	Electrostatics — Plates	61
	3.4	Electrostatics — Angled Plates	63
	3.5	Electrostatic Helmholtz	65
	3.6	Electric Quadrupoles	66
	3.7	Cylinder Boundaries	68
	3.8	Laplace Fourier	71
	3.9	Rotating Charged Sphere	73
	3.10	Dielectric Image Charge	75
	3.11	Dielectrics	78
	3.12	Dielectric Sphere	80
	3.13	Induction	82
	3.14	Magnetic Bottle	84
	3.15	A Fusion Reactor	86
	3.16	Dipole Radiation	90
	3.17	Two Antennas	92
	3.18	Dispersion	95
	3.19	Waveguide	96
	3.20	Skin Depth	99

4. Gases and Fluids   102

	4.1	Leaky Box	102
	4.2	Rectangular Flow	103
	4.3	Debye Temperature	105
	4.4	Times Arrow	108
	4.5	Compressibility	109
	4.6	Viscosity	112
	4.7	Water Waves	114
	4.8	Semiconductor	116

4.9	Semiconductor Junction	120
4.10	n-p Diode	124
4.11	Freezing Pipes	128
4.12	Cooling Earth — Interior	130
4.13	Cooling Earth — Exterior	132
4.14	Heat and pde	133
4.15	Heat Diffusion	135
4.16	Heat — Initial	137

5. Waves — 139

5.1	Longitudinal Slinky	139
5.2	Transverse Slinky	142
5.3	SR — Doppler	143
5.4	Step Response	145
5.5	Pulsating Sphere	146
5.6	Rectangular Drum	148
5.7	Piano	150
5.8	Aperture Diffraction Off Axis	150
5.9	Antenna	154
5.10	Antenna Array	156
5.11	Lissajous	156
5.12	Plane Waves — 2d	159
5.13	Damped, Driven Oscillator	160

6. Quantum Mechanics — 163

6.1	Box	163
6.2	Simple Harmonic Oscillator	163
6.3	Bound States — 3d Well	167
6.4	Identical Particles	170
6.5	Stark Effect	171
6.6	Square Well Scattering — 3d	173
6.7	Photoelectric Effect — Continuum	175
6.8	Line Width	179
6.9	Barriers	181
6.10	General Eigenvalues	185
6.11	Koenig–Penny	188

	6.12 Decay Chain	191
	6.13 Casimir Effect	192

7. Astrophysics 196

- 7.1 Transfer Orbit 196
- 7.2 Flyby 199
- 7.3 Lagrange Points 202
- 7.4 Binary Orbits 205
- 7.5 Three Body Orbits 207
- 7.6 Polytropic Star 209
- 7.7 Pulsating Stars 212
- 7.8 White Dwarf 215
- 7.9 Boltzmann and the Sun 218

8. General Relativity 222

- 8.1 Light Deflection 222
- 8.2 Circular Geodesic 223
- 8.3 General Geodesic 226
- 8.4 Kerr Photons 230
- 8.5 Kerr General 233
- 8.6 Gravity Wave Radiation 236
- 8.7 Stellar Pressure 238
- 8.8 Light Propagation in the Universe 244

Appendix: Scripts for the Chapter on Classical Mechanics 248

Graphical User Interface (GUI) 303

References 307

Index 309

Chapter 1

Mathematics

> "The scientist does not study nature because it is useful; he studies it because he delights in it, and he delights in it because it is beautiful. If nature were not beautiful, it would not be worth knowing, and if nature were not worth knowing, life would not be worth living."
> — **Jules-Henri Poincaré**

> "The chief forms of beauty are order and symmetry and definiteness, which the mathematical sciences demonstrate in a special degree."
> — **Aristotle**

1.1. Chaos

The world is a complex place and the relationship for complex systems between cause and effect is difficult to untangle. A famous statement about a fluttering butterfly wing "causing" a weather storm is an example of this perception. Complex systems have "chaotic" solutions. Predictions for macroscopic systems with 10^{23} particles or so by tracking each particle are clearly impossible. Even simple systems with few particles are not simply predictable if the initial conditions are changed very slightly.

A very simple system, consisting of a particle starting from a position and angle and bouncing around a totally rigid circular container is explored in the script "Chaos_Butterfly". Indeed, small changes in the initial conditions lead to large changes in the tracked trajectory after only twenty bounces. Plots for an initial position of y equal to zero and x equal to -0.5 (circular container of radius one) and initial angles of 20 and 21 degrees are shown in Figures 1.1 and 1.2 respectively. The patterns are quite different with a small change in the initial conditions. The user can "play" with different initial conditions and watch the subsequent behavior of the system in order to see how little the starting conditions need to be changed

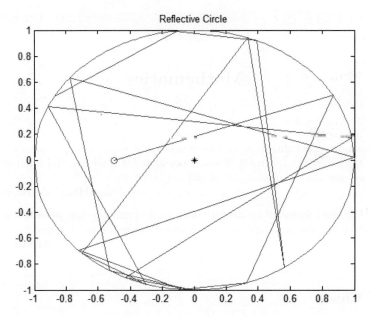

Figure 1.1: Trajectory for 20 bounces with an initial angle of 20 degrees. The starting point is the red circle, while the center of the circle is the black star.

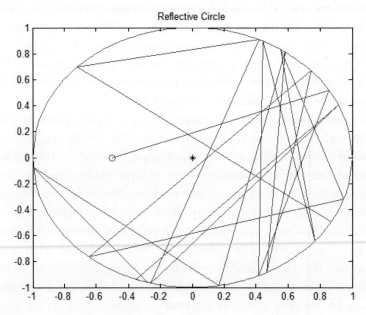

Figure 1.2: Trajectory for 20 bounces with an initial angle of 21 degrees. The starting point is the red circle, while the center of the circle is the black star.

to make a large change in the trajectory after only a few bounces off the container walls.

1.2. Malthus

Malthus long ago predicted that, with a fixed maximum food supply possible and with a growing population, famine would be unavoidable. However, so far, using fertilizers and crop improvements, this fate has been avoided by increasing the food supply. In the long run that approach has limitations which need to be considered and the predictions of Malthus will not be evaded forever.

The simplified equations explored in the script "Malthus" are:

$$\frac{dN}{dt} = aN$$
$$\frac{dc}{dt} = -aN + b \quad (1.1)$$
$$\frac{dN}{dt} = -\frac{b(N-c)N}{c}$$

The function N is the population as a function of time. If only the first equation in Equation (1.1) were active the number would grow exponentially with a characteristic time of $1/a$.

The "Malthus" equation assumes the number grows without limit, with exponential growth of the population which is a consequence of the first of Equation (1.1). A slightly more realistic model assumes the growth rate is time dependent and that rate, has a concentration of nutrients, c, which decreases with the population due to the spread of limited resources over an expanding population. The modified dN/dt value has two terms, bN/c which is growth with a time scale set by b/c and $-bN^2/c$ which decreases the growth with a rate which depends on the population N at any given time.

The symbolic solution to the last differential equation shown in Equation (1.1) is obtained using the utility "dsolve" in the script "Malthus", where the printout is shown in Figure 1.3.

Note that for late times the modified solution, Figure 1.3, approaches an asymptotic limited value, $N \to c$ and the population is self-limited by the availability of nutrients as the population grows

```
>> Malthus
   Bacterial Growth


   Bacterial Population - N, Limited Food Supply

      Nmo exp(a t)
      
                c
   ---------------------------
   c exp(-b t)
   ----------- - exp(-b t) + 1
      Nlo
   Enter the Growth Rate a (~1): 1
   Concentration of Nutrients b (0.3): 0.3
   Concentration Proportional to N, c (>0.5): 0.6
```

Figure 1.3: Solutions to the Malthus equation and modified equation, where Nmo is the initial population and Nlo is the initial population in the modified case which has limited resources.

in size. An example of the plots that can be formed by "Malthus" is shown in Figure 1.4, where the input parameters supplied by the user are $a = 1$, $b = 1$, $c = 5$. The reduction in the rate of growth with time is clearly evident as is the asymptotic approach of $N(t)$ to c, five in this specific example.

1.3. Missile Tracking

A mathematical problem arises when a missile acquires a target and moves to intercept it. The target is first seen at (x, y) of $(L, 0)$. The missile tracks to intercept by always aiming at the present location of the target at each moment. Presumably that position is instantly available by radar tracking, for example. The resulting differential equation to describe the missile trajectory, y as a function of x, is non-linear but solvable. The parameter q is the speed ratio of the missile to the target.

$$\frac{q(L-x)d^2y}{dx^2} = \sqrt{1 + \left(\frac{dy}{dx}\right)^2} \qquad (1.2)$$

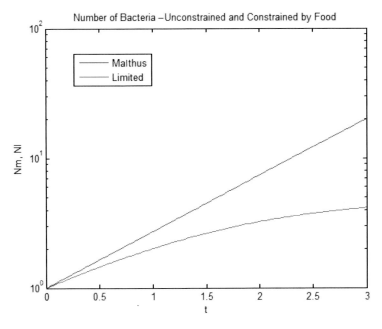

Figure 1.4: Plot of the time dependence of a population in the case of unconstrained resources, Malthus, and for limited resources. Both populations start at $t = 0$ with $N = 1$.

The script "Missile_Track" uses MATLAB "dsolve" to solve the equation symbolically. The actual form of the solution is not very transparent however. In the user dialogue, Figure 1.5, the input is the parameter n and the initial distance L when the target is first acquired. A movie of the pursuit is provided and the last frame for the specific case of L equal to ten and a speed ratio of two appears in Figure 1.6. The expression for the y coordinate of the intercept at x equal to L is $(L*q)/(q^2 - 1)$ which is plotted in Figure 1.7 where the distances are in units of L. The interception point appears in the symbolic solution, part of which appears in Figure 1.5.

1.4. Monte Carlo Analytic

The Monte Carlo technique is widely used in making models of complex systems. In the simplest view it consists of choosing variables

```
>> Missile_Track
  Solve for missile seeking target of slower speed


  Missile launched at (0,0) toward (L,0)
  Tracks by always aiming at present position of the target
  Missile velocity is q times the target velocity)
  Solved analytically

  ys =

  (L*q)/(q^2 - 1) - (q*(L - t)*(cosh(log(L - t)/q - log(L)/q)

  Enter x Location at y = 0 when target is acquired : 10
  Enter speed factor by which missile exceeds target : 2
  y of intercept = 6.66667
```

Figure 1.5: Printout of the user dialogue for "Missile_Track". The first term in the symbolic solution, ys, is the intercept point.

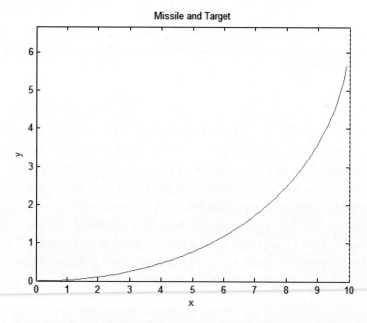

Figure 1.6: Last frame of a movie for the target, red line on the y axis, and the missile, blue line, intercept at y of 6.67.

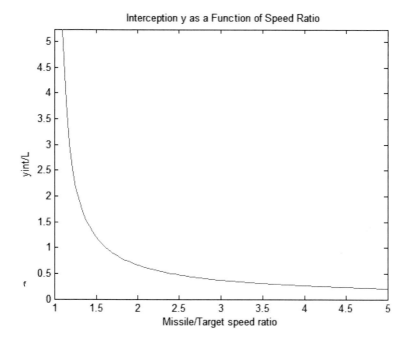

Figure 1.7: Plot of the y intercept in L units as a function of the speed ratio. The intercept diverges as the speed ratio approaches one, as expected.

from specified distributions, assembling a system from these choices and then studying the resulting performance of the system. In a few cases, the choice of a variable with a specific form of the distribution can be done analytically. Some of these cases are shown in the script "MC_Analytic" which uses the MATLAB random number generator "rand".

One solvable problem is the power law distribution of a variable x, where α is the desired power. The range of the variables is $x_{\min}$ to $x_{\max}$ and r is a random variable giving uniformly distributed values from zero to one.

$$\frac{\int_{x_{\min}}^{x} t^\alpha \, dt}{\int_{x_{\min}}^{x_{\max}} t^\alpha \, dt} = r$$

$$x = \left[x_{\min}^{\alpha+1} + r \left(x_{\max}^{\alpha+1} - x_{\min}^{\alpha+1} \right) \right]^{1/(\alpha+1)} \quad (1.3)$$

A second example is an exponential distribution with a lifetime τ.

$$\frac{\int_{x_{\min}}^{x} e^{-t/\tau}\,dt}{\int_{x_{\min}}^{x_{\max}} e^{-t/\tau}\,dt} = r$$

$$x = -\tau \ln[e^{-x_{\min}/\tau} + r(e^{-x_{\max}/\tau} - e^{-x_{\min}/\tau})] \quad (1.4)$$

The Brite-Wigner resonance line shape can also be solved analytically and is shown in Equation (1.5). The central value is x_o and the width, full width at half maximum is Γ.

$$\frac{\int_{x_{\min}}^{x}\left[\frac{1}{(t-x_o)^2+(\Gamma/2)^2}\right]dt}{\int_{x_{\min}}^{x_{\max}}\left[\frac{1}{(t-x_o)^2+(\Gamma/2)^2}\right]dt}$$

$$\varphi_{\min} = \frac{2(x_{\min} - x_o)}{\Gamma} \quad (1.5)$$

$$\varphi_{\max} = \frac{2(x_{\max} - x_o)}{\Gamma}$$

$$x = \frac{x_o + \Gamma}{2\tan\{\tan^{-1}(\varphi_{\min}) + r[\tan^{-1}(\varphi_{\max}) - \tan^{-1}(\varphi_{\min})]\}}$$

A Gaussian distribution can be achieved by using two random numbers and the fact that the joint probability of two uncorrelated variables is the product of the probabilities. In Equation (1.6) the standard deviation of the Gaussian is σ. The joint probability can have a "radius" chosen from an exponential and an azimuthal angle chosen randomly from zero to 2π. The means of the Gaussian can simply be added to the resulting x and y values.

$$\begin{aligned}
d\underline{P}(x)\,d\underline{P}(y) &= e^{-x^2/2\sigma^2} e^{-y^2/2\sigma^2}\,dx\,dy \\
&= e^{-r^2/2\sigma^2} r\,dr\,d\varphi \\
&= e^{-u/2\sigma^2}\frac{du\,d\varphi}{2}, \quad u = r^2 \quad (1.6)
\end{aligned}$$

The script, "MC_Analytic" gives the user a choice of these distributions and the parameters which specify the distributions, such as the desired power where the menu presented by the script appears in Figure 1.8. An example printout for a resonance shape appears in

Figure 1.8: Menu choices presented by the script, "MC_Analytic".

```
>> MC_Analytic
   Monte Carlo - analytic method of inversion/integration


   Monte Carlo - Analytic Examples
    Enter Mean: 5
    Enter Linewidth: 2
    Enter x Minimum: 0
    Enter x Maximum: 10
```

Figure 1.9: Example of the dialogue for the choice of a resonance distribution.

Figure 1.9 and the output is a histogram of the distribution, displayed in Figure 1.10.

In this script the MATLAB utility "hist" is used to create histograms of the numerical results. The options available can easily be seen in the Command Window by typing in "help hist".

1.5. Monte Carlo Numeric

In the vast majority of cases the integral which is needed in the analytic case either cannot be done analytically or the result cannot be simply inverted. In such cases a variety of numeric techniques are available. The simplest is to choose two random numbers. The first, r_1, uniformly populates the x range desired, while the second, r_2, weights the chosen x by the probability distribution $P(x)$. The user

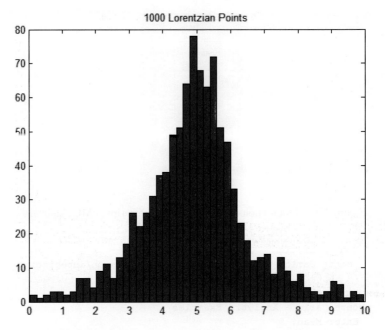

Figure 1.10: Example of the output plot for the choice of a resonance distribution with 1000 generated entries.

is encouraged to make a graph of the procedure which will show why it works. It is here assumed that $P(x)$ has a maximum in the x range being used. Using the maximum makes the choices of the random numbers more efficient. The procedure is shown in Equation (1.7).

$$\begin{aligned} &\text{pick } r_1 := x_{\min} + r_1(x_{\max} - x_{\min}) \\ &\text{pick } r_2 : \quad \text{if } \frac{r_2 < P(x)}{P(x)_{\max}} \quad \text{then accept } x \\ &\phantom{\text{pick } r_2 :} \quad \text{if } \frac{r_2 > P(x)}{P(x)_{\max}} \quad \text{then reject } x \end{aligned} \qquad (1.7)$$

An example from the script "MC_Numeric" appears in Figure 1.11. In this case the angular distribution for Compton scattering with user defined initial photon energy is produced. The user can vary

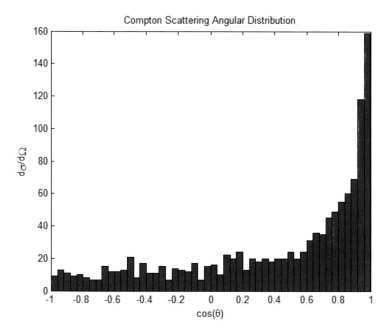

Figure 1.11: Angular distribution in Compton scattering for a specific, user defined, photon energy.

the energy and see how the distribution varies, as desired. This distribution is complex and cannot be found analytically.

1.6. Monte Carlo Muon Beam

As an example of a somewhat more complex modeling problem, an initial beam of muons is defined by the user and the particles in the beam are tracked through a block of iron where they lose energy by ionization and also multiple scatter in the material. The script is called "Monte_Muon". This package attempts to give the user a picture of how a complex system might be modelled and then tracked through a particular environment.

The script uses the functions "Gauss", "PowerLaw", "Mult_Scatt" and "Euler." The first two functions were described above and provide Gaussian distributed or power law distributed variables. The function "Euler" transforms a momentum vector, p, from a frame with the z axis oriented along spherical angles ϑ and φ to the "laboratory" frame, p', where the z axis has those angles. A check for the special case where the momentum is along the z axis, $p_x = p_y = 0$, yields the momentum components in the primed system which are as expected, seen as the third column of the transformation matrix.

$$\begin{bmatrix} p'_x \\ p'_y \\ p'_z \end{bmatrix} = \begin{bmatrix} \cos\theta\cos\phi & \sin\phi & \sin\theta\cos\phi \\ \cos\theta\sin\phi & \cos\phi & \sin\theta\sin\phi \\ -\sin\theta & 0 & \cos\theta \end{bmatrix} \begin{bmatrix} p_x \\ p_y \\ p_z \end{bmatrix} \quad (1.8)$$

The Euler angles define a slightly more general transformation and they will be defined and used in the next chapter of this text. They relate the frame attached to a solid body to a frame related, in general, by three angles defining the orientation of an extended body.

The script "Mult_Scatt" uses "Euler" and "Gauss".

A muon is a charged particle with a rest mass of 106 MeV. It traverses a length of material and suffers a scattering angle ϑ by an amount chosen from a Gaussian distribution with a defined mean of zero and a standard deviation fixed by the length of material being travelled and by the muon momentum. Because the angles of scattering are not small at low energies or for thick material and the energies may change substantially, the block of material to be traversed is divided into segments, the number of which is chosen by the user. The segments should be such that a substantial fraction of the muon energy is not lost in any single segment nor should the scattering angles be large. A specific dialogue is shown in Figure 1.12.

The beam starts at normal incidence to the block of iron. Assuming a path strictly normal the beam would lose 8 MeV in energy and scatter by 38 degrees after traversing the full 50 cm of iron. However, the path is more complex than that, which is what necessitates the more detailed Monte Carlo modelling.

1. Mathematics

```
>> Monte_Muon
  Program to make a model of a muon beam passing through a block of Fe


Radiation Length, min dE/dx, Density and Muon Mass
Enter Mean Initial Muon Momentum (MeV), dp/p = 3%: 200
Enter Transverse sigma (cm): 0.1
Enter z Length of Block (cm): 50
Enter Number of Numerical Strips for Block of Fe: 50
Muon at the End of range
Muon at the End of range
Muon at the End of range
Muon at the End of range
Muon at the End of range
Muon at the End of range
Muon at the End of range
Muon at the End of range
Initial Transverse Size =0.0811707
Final Transverse Size =15.4576
Initial Energy =226.116
Final Energy =212.665
Initial Mean Momentum =199.922
Final Mean Momentum =184.5
```

Figure 1.12: Specific choice of the muon beam of 200 MeV mean momentum.

The 200 MeV mean momentum beam loses about 15 MeV on average and about 14 MeV in energy on average. In addition, the multiple scattering may lead to long distances in the iron block, so that "End of Range" or the loss of all kinetic energy of a given muon must be protected against. The energy loss is highly dependent on the muon velocity, going as the inverse square, so that behavior should be modelled properly.

The starting energy distribution is shown in Figure 1.13, while the exiting distribution appears in Figure 1.14. Note the long, low energy tail due to the traversal of many cm of iron. As before, the utility "hist" is used to make the plots.

There is a long, low energy tail which arises from multiple scattering in the block, whose transverse size is assumed to be very

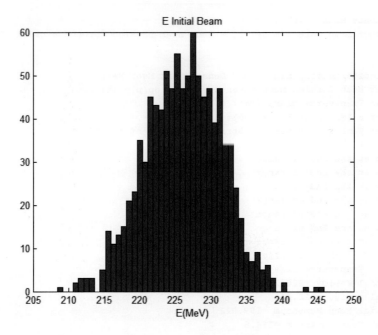

Figure 1.13: Distribution in initial energy for a 200 MeV momentum mean muon beam with a 3% momentum spread.

large. The actual path length for the muon with large scattering angles then becomes very long. The initial energy was confined to the region (205,250) MeV as seen in Figure 1.13.

The x and y locations of beam particles for the initial beam and final beam are shown in Figure 1.15. Note the scale change in the two plots by a factor of 20. These two plots underscore the need to do correct modelling. In particular, note that a lower energy muon has more multiple scattering for a given thickness of material.

1.7. Fourier and Laplace Transforms

Fourier and Laplace transforms are useful quantities in several applications in physics. The Fourier transform is defined in Equation (1.9). The inverse has the same definition save that the argument of the exponential changes sign. For the Laplace transform, the inverse, $f_{\text{ILT}}(x)$, has a different integration contour, one where the

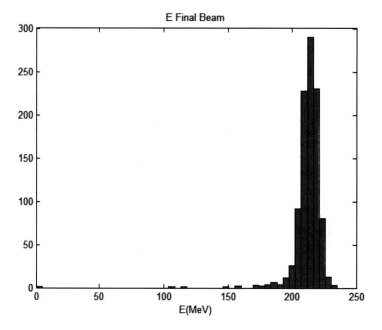

Figure 1.14: Distribution in final energy for a 200 MeV muon beam momentum with 3% momentum spread having traversed a 50 cm block of iron. The entries at zero energy have, in fact, stopped completely.

singularities of $F(s)$ are assumed to be to the left of the parameter a.

$$F_{\text{FT}}(s) = \int_{-\infty}^{\infty} f(x) e^{-2\pi i x s} \, dx$$

$$F_{\text{LT}}(s) = \int_{0}^{\infty} f(x) e^{-xs} \, dx \tag{1.9}$$

$$f_{\text{ILT}}(x) = \left(\frac{1}{2\pi i}\right) \int_{a-i\infty}^{a+i\infty} F(s) e^{xs} \, ds$$

The transforms can be explored using the script "Fourier_Laplace_Trans." The dialogue first asks if another transform is desired and then asks that Fourier of Laplace be specified. Then, the user specifies whether the transform or the inverse is desired. The printout of a dialogue to find the Laplace transform of a specific function appears in Figure 1.16. The integrals often appear with

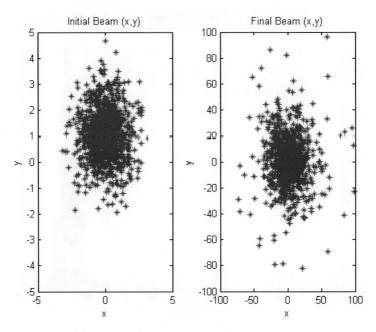

Figure 1.15: Distributions of muon positions in the beam at the start of the traversal of the iron (left), and after complete traversal of the iron block (right). The scales are different by a factor of 20.

```
>> Fourier_Laplace_Trans
   Solve symbolically for the FT or LT of a function


   Find continuum FT/LT for user defined function
Input the function, f(x)
: sin(x)

           /    exp(-x y) (cos(x) + y sin(x))         \         1
    limit  | -  -----------------------------, x = Inf | + ------
           |                  2                        |       2
           \                 y + 1                     /      y + 1
Input the function, f(y) for plots
: 1/(y^2+1)
Enter range of x (-xl,xl): 3
Enter range of y (-yl,yl): 3
```

Figure 1.16: Example of the Laplace transform for $\sin(x)$. The user must input the correct limit in order to invoke plots of $f(x)$ and $F(y)$.

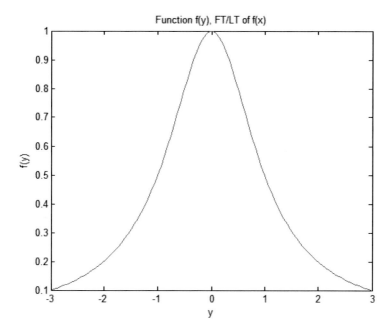

Figure 1.17: Laplace transform, $F(y)$ of $f(x) = \sin(x)$.

limits because they are found symbolically using the MATLAB utility "int". When that happens the user is asked to supply the correct result, as seen in Figure 1.16 for example, and then plots of $f(x)$ and $F(y)$ are created. The specific plot appears in Figure 1.17.

The Fourier transform for an exponential and a Gaussian can be obtained. The Laplace transform for a power of x, square root of x and $1/x$, exponential I x, power of x times an exponential in x, and $\sin(x)$ or $\cos(x)$ all give correct results. The inverse Laplace transforms are less successful, but the user can obtain the pairs for the tranforms by asking for the Laplace transform itself. These are continuum results. In the discrete case the MATLAB utilities "fft" and "ifft" yield the desired results.

Chapter 2

Classical Mechanics

"The most exciting phrase to hear in science, the one that heralds new discoveries, is not 'Eureka!' but 'That's funny' ..."
— **Isaac Asimov**

"Measure what can be measured, and make measureable what cannot be measured."
— **Galileo Galilei**

2.1. Angular Momentum

Angular momentum, L, is fundamental to many aspects of physics. For example, since the vector L is the cross product of the position and the momentum, it is a conserved quantity for central forces. These forces cause orbits confined to a plane perpendicular to L, as for example in the Kepler planetary orbit problem. This constant of the motion allows the motion in the plane to be reduced to a one-dimensional problem for central forces.

$$\frac{d\vec{L}}{dt} = \frac{d(\vec{r}xm\vec{v})}{dt} = m(\vec{v}x\vec{v} + \vec{r}x\vec{a}) = 0 \qquad (2.1)$$

Two-dimensional motion in the (x, y) plane is assumed in the script "Angular_Momentum". The solution is symbolic using the user supplied symbolic trajectories $x(t)$ and $y(t)$ to compute $L(t)$. A specific printout is displayed in Figure 2.1.

A movie of the time development of the trajectory and the Lz component is then made. In the specific example the plot is shown in Figure 2.2. Note that Lz is not here a constant of the motion because the trajectory does not correspond to a central force result. The user should try several symbolic trial functions and see how Lz evolves.

2. Classical Mechanics

```
>> Angular_Momentum
   Symbolic Angular Momentum


   Find L for Symbolic x(t), y(t) Input
   Enter x(t) symbolic
   : t^3
   Enter y(t) symbolic
   : cosh(t)
   Angular momentum (z-component) =

   Lz =

   t^3*sinh(t)  -  3*t^2*cosh(t)
```

Figure 2.1: Printout from the script "Angular_Momentum" where L_z is calculated symbolically given the user input for $x(t)$ and $y(t)$.

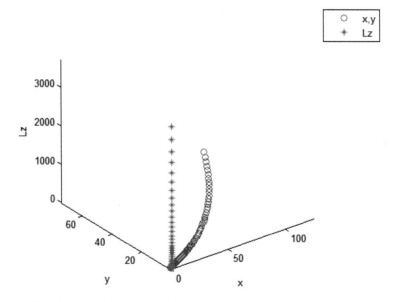

Figure 2.2: Time development of the (x, y) trajectory and the angular momentum. Note that L_z is not a constant on the trajectory.

2.2. Classical Scattering

Scattering off a force center is used classically to explore the forces between the target and the scattered particle. However, the total scattering cross section is not unique to a given force law. An example is the three cases which give the same total cross section πa^2; a hard sphere of radius a, a barrier of height V_o and width a', and a well of depth V_o and width a. For a hard sphere, the differential cross section, $d\sigma = 2\pi b\, db$, with no scattering for radii greater than a, yields the cross section $\sigma = \int_0^a (d\sigma/db)db = \pi a^2$.

The relationships between the impact parameter, b, and the scattering angle, θ, for the three cases are given below. For the hard sphere the angle is zero when b is a and π when b is zero.

$$\theta_{hs} = \pi - 2\sin^{-1}\left(\frac{b}{a}\right), \quad b < a$$

$$= 0, \quad b > a$$

$$\theta_{bar} = 2\left(\sin^{-1}\left(\frac{b}{na'}\right) - \sin^{-1}\left(\frac{b}{a'}\right)\right), \quad b < b_o \qquad (2.2)$$

$$= \pi - \sin^{-1}\left(\frac{b}{a'}\right), \quad b > b_o, \quad n^2 = \frac{1-V_o}{E}, \quad b_o = na'$$

$$\theta_{well} = -2\left(\sin^{-1}\left(\frac{b}{na}\right) - \sin^{-1}\left(\frac{b}{a}\right)\right), \quad n^2 = \frac{1+V_o}{E}$$

A plot of the three scattering angles with energy E equal to twice the magnitude of the potential, where appropriate, as a function of impact parameter is made in the script "Classical_Scatt" and shown in Figure 2.3. It appears that they are rather different. However, since the differential cross section goes as $b\,db$, the total cross section contributions are very heavily weighted toward large b, and the small b scattering contributes little to the total cross section. The conclusion of the plot is that one must measure the differential cross section in order to make a more incisive determination of the force law which is active in the scattering.

2.3. Impact Parameter and DCA

The impact parameter, b, is the perpendicular distance between an incident particle direction and the force center location at large initial separations. The distance of closest approach, a, is that point on the

2. Classical Mechanics

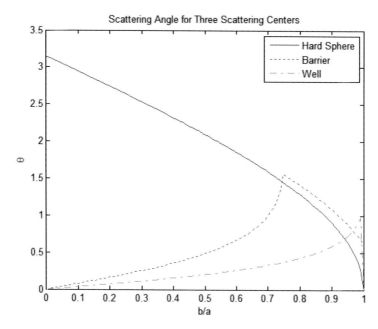

Figure 2.3: Scattering angle vs. impact parameter for the three classical interactions. The specific choice for the barrier has a' larger than a by 5%.

particle trajectory closest to the force center, which is assumed to be fixed. They are related by applying the conservation of energy and the conservation of angular momentum for a general power law central force as shown in Equation (2.3).

$$
\begin{gathered}
f(r) \sim \frac{m}{r^\alpha}, \quad V(r) \sim \frac{m}{(\alpha-1)r^{(\alpha-1)}} \\
v_o b = v a \\
v_o^2 = v^2 + \frac{2V(r)}{m}, \quad v_o^2 = \frac{2E}{m} \\
a^2 \left(\frac{1 - V(a)}{E} \right) = b^2
\end{gathered}
\tag{2.3}
$$

Printout from the script "CM_b_DCA" is shown in Figure 2.4 for $\alpha = 2$, attractive and repulsive and 3 for an attractive force. The script uses the MATLAB utility "solve" to find an explicit closed form solution if one exists.

```
>> CM_b_DCA
Program to look at the relationship of impact parameter
and distance of closest approach - force power laws


Particle with impact parameter b scatters off a force center
a = vo*b/v, a is DCA, initial velocity vo, v at DCA
f(r) = m /r^a, V(r) = -m/(a-1)r^a-1
E conservation; vo = sqrt(v ^2 + 2 /((alf-1) *(r ^(alf-1))))
Enter force law power: 2

aa =

 ((b^2*vo^4 + 1)^(1/2) - 1)/vo^2
-((b^2*vo^4 + 1)^(1/2) + 1)/vo^2

Enter force law power: 2

aa =

 ((b^2*vo^4 + 1)^(1/2) + 1)/vo^2
-((b^2*vo^4 + 1)^(1/2) - 1)/vo^2

Enter force law power: 3

aa =

(b^2 + 1/vo^2)^(1/2)
```

Figure 2.4: Printout of "CM_b_DCA" for inverse power law of 2, repulsive and attractive and power of 3, repulsive.

Other power laws are possible and the user can try them. Plots of the ratio of a/b for the case of an inverse cubic power law force as a function of the incident velocity are shown in Figures 2.5 and 2.6 for repulsive and attractive forces respectively. In the case of a repulsive force the ratio a/b is always greater than one and only approaches one at high incident velocity. The opposite behavior is seen for attractive force. At low velocity the projectile is pulled to the force center and small a/b values are seen. Only at large velocity does the point of closest approach reach the limit of the impact parameter.

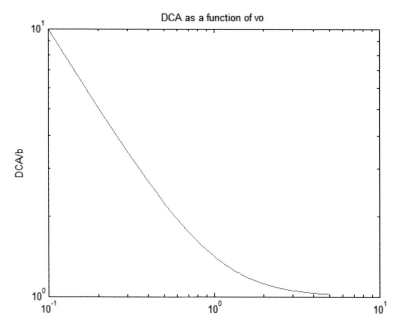

Figure 2.5: Ratio of a/b as a function of incident velocity (arb. units) for a repulsive force. At low velocity the particle is repelled to large values of a/b, while at high velocity the power 3 repulsive force is relatively weaker and $a > b$.

2.4. Foucault Pendulum

A Foucault pendulum is one installed on the Earth's surface which thus exists in an accelerated coordinate system due to the Earth's rotation. One possible way to think about it is that the pendulum oscillates in "absolute space" while the Earth rotates under it. In the Earth frame the equations of motion have an added term due to the Coriolis force, where Ω is the circular frequency of the Earth and φ is the latitude. In the absence of rotation the equations of motion describe simple harmonic motion in the (x, y) plane of the Earth's surface. With the rotation, the motion in x and y is coupled due to the terms proportional to velocity. In Eq. (2.4) the symbol ω refers to the frequency of the pendulum in the absence of

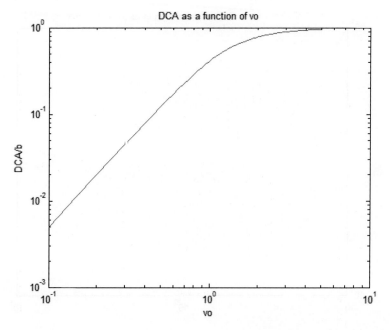

Figure 2.6: Ratio of a/b as a function of incident velocity (arb. units). At low velocity the particle is attracted to the force center and small values of a/b are observed, while at high velocity the power 3 attractive force is relatively weaker and $a > b$.

accelerated motion.

$$\frac{d^2x}{dt^2} = -w^2 x + 2\Omega \sin\phi \left(\frac{dy}{dt}\right)$$
$$\frac{d^2y}{dt^2} = -w^2 y - 2\Omega \sin\phi \left(\frac{dx}{dt}\right)$$
(2.4)

These equations are solved symbolically using the "dsolve" MATLAB utility in the script "Foucault_exact". The approximate solution for the case where ω, the pendulum harmonic frequency, is much less than Ω is also shown for comparison. The script is quite slow because at least an entire day is mapped out; be patient. The exact x and y positions in the Earth frame are shown in Figure 2.7 for the case where the pendulum period is 0.05 days and it sits at thirty degrees

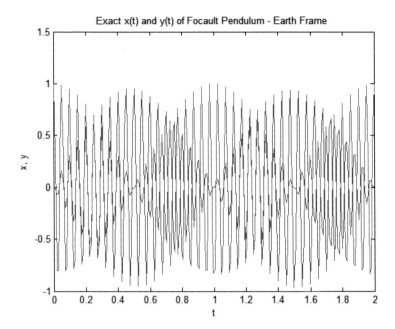

Figure 2.7: Plot of x and y locations of a Foucault pendulum as a function of time in days. These trajectories are re-entrant in time with period of 1 day.

north latitude. The approximate solution is also plotted by the script as a movie and illustrates that the motion is not re-entrant on a daily basis in this case because of the approximations made.

The approximate solution factorizes into simple harmonic motion times a modulation of the amplitude due to the Coriolis force.

$$\Omega' = \Omega \sin \phi$$
$$x = \sin(\Omega' t) \sin(\omega t) \tag{2.5}$$
$$y = \cos(\Omega' t) \sin(\omega t)$$

The plot in Figure 2.8 shows the exact solution in the pendulum frame in blue and a view in the "absolute" frame where the pendulum exhibits simple harmonic motion. The movie of the two motions is plotted and Figure 2.8 is the last frame of that movie.

A plot of the (x, y) motion of the pendulum (exact) is shown in Figure 2.9 for the case where the pendulum period is 0.1 days at

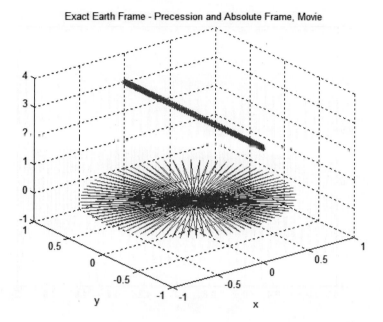

Figure 2.8: Plot of the pendulum position, in the exact case, on Earth (blue) and in absolute space (red stars) as the last frame of a movie of the Foucault motion.

30°N latitude. The motion is clearly nicely re-entrant after one day of motion.

2.5. Spring Pendulum

The simple pendulum problem can be extended by making the mass support have spring properties. The equations of motion are then:

$$\begin{aligned}\frac{d^2 r}{dt^2} &= g\cos\theta + \left(\frac{k}{m}\right)(L_o - r) + r\left(\frac{d\theta}{dt}\right)^2 \\ \frac{d^2\theta}{dt^2} &= -\frac{g\sin\theta}{r} - 2\frac{\left(\frac{dr}{dt}\right)\left(\frac{d\theta}{dt}\right)}{r}\end{aligned} \qquad (2.6)$$

where r is along the spring with coefficient k and un-stretched length L_o and the angle θ is the angle with respect to the vertical axis. The results of a user dialogue with the script "Spring_Pendulum" which uses the MATLAB script "ode45" are shown in Figure 2.10.

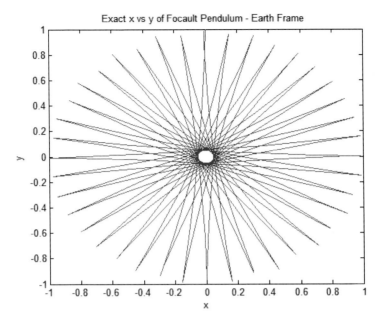

Figure 2.9: Plot of the pendulum position, in the exact case, on Earth, as it evolves in time. The pendulum period is 0.1 days. The trajectory is plotted for 2 days by which time the pendulum has returned to the initial configuration.

```
>> Spring_Pendulum
Enter k, m and Lo - [k,m,Lo]: [3 1 10]
Spring contributes as (Lo-r), >0 and <0
w for simple spring = 1.73205 and simple pendulum = 0.989949
Enter Initial Positions , r and theta(deg) - [r(0),theta(0)]: [12 30]
Initial Velocities  Assumed to be Zero
```

Figure 2.10: Dialogue for "Spring_Pendulum" defining the spring and the pendulum aspects of the problem.

The spring elongates and compresses and the mass point follows while simultaneously being pulled toward the vertical by gravity. The resulting radius as a function of time with respect to the un-stretched spring at L_o is displayed in Figure 2.10. A movie is provided by the script and the last frame of the movie is shown in Figure 2.11.

The user can check the solutions by going to the limiting cases of g equal to zero, the spring alone, and the case where k is equal to zero, the simple pendulum. In general the motion is not simple harmonic.

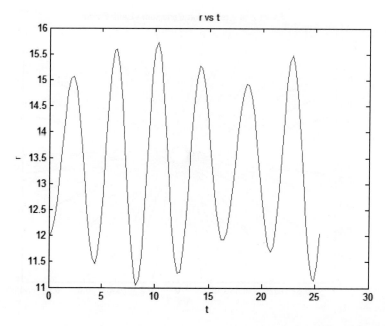

Figure 2.11: Plot of the radius as a function of time. The unstretched length is 10 in this example, where the parameters chosen appear in Figure 2.10, where the initial r is 12. Changes in r indicate stretching or compression of the spring which supports the mass point.

For example, the plot in Figure 2.11 displays a fairly complex pattern where the maximum amplitude in each oscillation period changes as does the frequency itself. The trajectory is illustrated in the last frame of a movie for the (x, y) motion of the mass point for a slightly different set of parameters in Figure 2.12. The initial velocities are assumed to be zero in this script.

2.6. Spherical Pendulum

The spherical pendulum is examined in the script "Spherical_Pend" which employs the MATLAB tool "ode45" to provide a numerical solution to the problem. The pendulum has a mass point on a fixed length r_0 with gravity, g, acting on the mass point. The mass point moves on the surface of a sphere and there is a conserved angular momentum so that the problem is reduced to one dimension. The

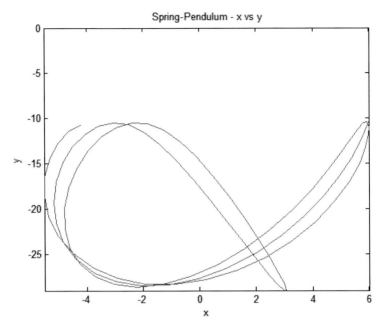

Figure 2.12: Last frame of a movie of the motion of the spring pendulum for parameters defined to be $[kmL_o] = [1\ 1\ 10]$ and initial r and angle $= [12\ 50']$.

equations of motion for the spherical coordinates are:

$$\frac{d^2\theta}{dt^2} - \frac{(L^2 \cos\theta)}{\sin^3\theta} - \left(\frac{g}{r_0}\right)\sin\theta = 0$$
$$\frac{d^2\phi}{dt^2} + 2\frac{\left(\left(\frac{d\theta}{dt}\right)L\cos\theta\right)}{\sin^3\theta} = 0 \quad (2.7)$$

As can be seen in Eq. (2.7) the polar angle θ has acceleration terms due to the centrifugal force, the L2 term, and due to gravity, the g term, pulling the mass term toward the vertical. The azimuthal angle term simply expresses the conservation of the angular momentum L, and is included to easily evaluate the results for that angle.

The script asks for $\frac{g}{r_o}$, L and the initial angle and initial angular velocity. After defining the initial conditions a movie of the subsequent motion is provided and plots for a specific case are shown in Figures 2.13 and 2.14.

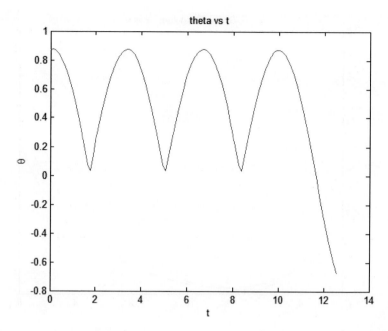

Figure 2.13: Plot of the polar angle as a function of time for the specific case where $g/r_o = 1, L = 0.01, \theta = 50°$ and $d\theta(0)/dt = 4$ degrees/sec.

The user is encouraged to try a full range of possible initial conditions and a range of parameters defining the spherical pendulum. For example, another set of parameters and initial conditions are the input to the plot of Figure 2.15.

2.7. Driven Pendulum

A system that displays chaotic behavior is the driven pendulum. The script "Driven_Pendul" allows for an exploration of this behavior. The pendulum is damped and harmonically driven, so that the equation of motion is:

$$\frac{d^2\theta}{dt^2} + \frac{\Gamma d\theta}{dt} + \omega_o^2 \sin\theta = F_o \cos(\omega_d t) \qquad (2.8)$$

The pendulum moves in one dimension only. The damping factor is Γ, the amplitude of the driving force is F_o and the driving frequency is ω_d while the natural frequency ω_o is taken to be one in the script.

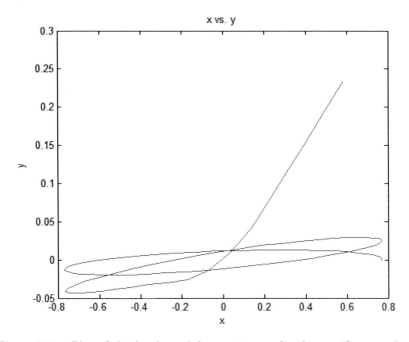

Figure 2.14: Plot of the (x, y) pendulum trajectory for the specific case where $g/r_0 = 1, L = 0.01, \theta = 50°$ and $d\theta(0)/dt = 4$ degrees/sec.

There is a region of operation where the motion is largely harmonic. In the case where the initial angle is fifty degrees, initial angular velocity is zero, damping is 0.5, F_o is 0.2 and ω_d is 0.6 the motion is shown in Figure 2.16. A movie is also provided by the script. If the damping is changed to 0.2, the amplitude to 1.5 and the frequency moved to 0.7, the behavior changes and the pendulum loops over the horizontal as seen in Figure 2.17. Each problem may have a region of validity governed by the parameters of the problem and the user can explore over which region of parameters the characteristic behavior occurs.

In the quasi-harmonic regime, Figure 2.16, the pendulum itself has transient time structure which dies out so that at long times the behavior is controlled by the driving term frequency. In the chaotic regime, the connection between the response and the driving harmonic function is less obvious.

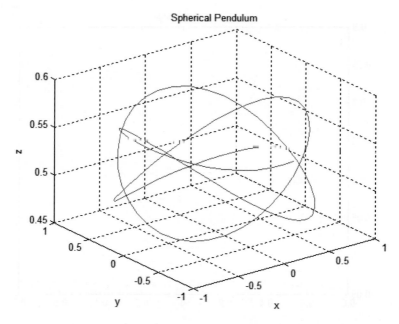

Figure 2.15: Plot of the (x, y, z) pendulum trajectory for the specific case where $g/r_0 = 1, L = 1, \theta = 60°$ and $d\theta(0)/dt = 5$ degrees/sec.

2.8. Random Walk

The random walk refers to steps taken in a random direction. This is the "drunkards walk" problem; if a drunkard starts out from a lamppost and takes 100 steps, how far does he get from the lamppost? The path taken is displayed using the script "rand_walk". The user chooses the number of steps and subsequently a movie of the path of those steps in made. In this specific case the walk is in two dimensions and the length of each step is one unit. The walk starts from the origin.

A plot for one hundred steps is shown in Figure 2.18. This is clearly a stochastic problem with a random path. A physical example of such a path is the Brownian motion of a particle scattered randomly by the fluid in which it is immersed. The distance traveled is not zero, but goes as the square root of the number of steps. This behavior is characteristic of stochastic processes. A plot of

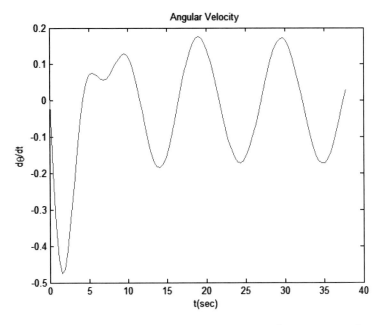

Figure 2.16: Angular velocity in the stable regime of parameters where the motion is quasi-harmonic.

the radius of the walk for one hundred random steps is displayed in Figure 2.19.

Note that the radius is not zero, but does display the square root dependence which is expected. For one hundred steps a mean radius of ten might be expected and is approximately observed. Another example of stochastic behavior is that exhibited in multiple Coulomb scattering where the mean overall scattered angle goes as the square root of the distance travelled by the charged particle. This was assumed in the muon beam exercise previously considered as coded explicitly in the script "Mult_Scatt" where the mean multiple scattering angle goes as the square root of the material distance travelled in radiation length units.

2.9. Rotating Hoop

The normal pendulum problem is set up in an inertial system and displays simple harmonic motion for small oscillations about the

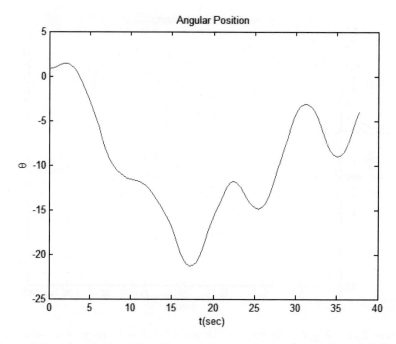

Figure 2.17: Angular position for parameters appropriate to chaotic motion. Note that the range of angles is greatly expanded showing that the pendulum has crossed the horizontal several times during the time period set for the display.

vertical equilibrium position. An extension of this situation looks at a hoop, radius a, with a sliding mass point, m, on it, where the hoop rotates with a circular frequency ω. There are two equilibrium positions possible, one at the bottom vertical of the hoop and one at a non-zero angle with respect to the vertical.

The equations of motion follow from the Lagrangian formed from the potential, V, and two contributions to the kinetic energy, T. The first term in the expression for T is due to the mass point velocity. The second term in T is due to the rotation of the hoop due to the centrifugal effect.

$$T = m\frac{\left(a\frac{d\theta}{dt}\right)^2}{2} + m\omega^2\frac{(a\sin\theta)^2}{2}$$
$$V = -mga\cos\theta \tag{2.9}$$

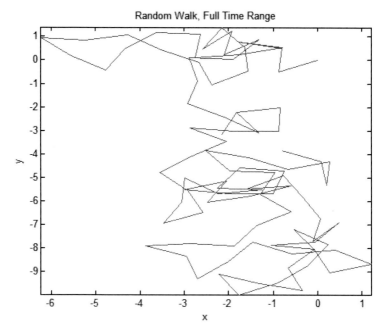

Figure 2.18: The path of a random walk in x and y with 100 steps of length one taken.

Using the Lagrangian, $L = T - V$, and the Lagrange equations of motion, the resulting angular acceleration is zero if $\omega = \sqrt{g/a \cos\theta}$. The critical frequency squared for the second equilibrium position to exist is g/a and harmonic motion about the second equilibrium position occurs for small angles with respect to that position with a frequency $\Omega = \omega\sqrt{1 - (g/a\omega^2)^2}$.

The script "Rotating_Hoop2" is used to solve this problem numerically using the MATLAB utility "ode45". The radius and mass are taken to be equal to one. A specific example, with frequency four, no initial velocity for the mass, and an initial angle with respect to the second equilibrium position of 0.3 radians is displayed below. Printout of the dialogue appears in Figure 2.20.

A movie of the motion of the hoop and mass point is displayed, showing the approximately harmonic motion of the mass point for this choice of parameters. The last frame of the movie appears in

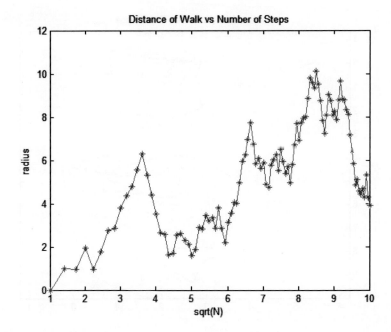

Figure 2.19: The radius from the starting point of a random walk with 100 steps of length one taken.

```
Hoop Rotates with frequency w, w^2 > g/a for equilibrium not equal to 0
 Enter Frequency of Oscillation, > sqrt(g) (~4): 4
Stable Angle for Small Oscillations = 0.911577 (rad)
Enter Initial Angle w.r.t. Stable Angle (rad): 0.3
Enter Initial Velocity (rad/sec): 0
```

Figure 2.20: Printout of the script "Rotating_Hoop2" for a user supplied input of the hoop rotation frequency. In this case the stable angle is 0.91 radians.

Figure 2.21. The user can create several movies in order to get a feeling for the parameter dependence of the problem.

2.10. Tides

Newton first explained the tides. As was understood much later, it is the tidal forces that are important in General Relativity because any overall acceleration can be removed by going to a free falling frame as stated by the Equivalence Principle. The tidal forces are

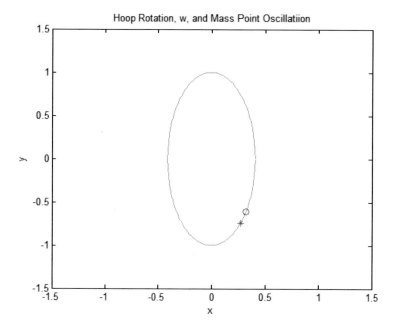

Figure 2.21: Frame of a movie of the rotating hoop. The hoop is drawn in green, the mass point in red, and the equilibrium angle is a blue o. The hoop rotates in the movie in order to show the relationship of the hoop rotation to the mass point oscillation.

defined locally at a mass point. They both stretch and compress an extended body.

Due to the attraction of the moon there are tidal forces which are in directions and of magnitude such as to elongate the oceans in a line with the moon. For the moon-earth distance of r_0, the forces F_z and F_x at a location specified at a point (x, z) near radius r_0 with z along the moon-earth direction and x transverse to z are:

$$F_z = -GM \left[\frac{1}{(z+r_0)^2} - \frac{1}{r_0^2} \right] \sim \frac{2zGM}{r_0^3}$$

$$F_x \sim -\frac{xGM}{r_0^3} \qquad (2.10)$$

$$V(x,z) = -\frac{GM}{r_0^3} \left[z^2 - \frac{(x^2+y^2)}{2} \right]$$

The forces can be integrated to yield the gravitational potential, $V(x,z)$ describing the tidal forces.

The additional forces cause the tides to rise to a height h, whose potential gh balances the potential added by the moon where the tides are assumed to be equipotentials of the system. The tidal height, low to high, is about, $\Delta h \sim 0.6\,\text{m}$ in this approximation and depends on the masses of the Earth and moon and their radii, M_m, M_e, R_m, R_e. There are clearly two tides per day. The angle θ defines the angle between a point fixed on the earth at radius r and the location of the moon in its orbit. In polar coordinates the potential is:

$$r = \sqrt{x^2 + z^2}$$
$$V(r,\theta) = -\left(\frac{GM}{2r_0^3}\right)(3\cos^2\theta - 1)r^2$$
$$gh - V(r,\theta) = const \quad (2.11)$$
$$\Delta h = \frac{3GMr^2}{(2gr_0^3)}$$

A movie of the tides is provided by the script "Tides". A frame of that movie appears in Figure 2.22, where the spherical earth is red, while the ocean tides are in dotted blue. The location of the moon is a blue circle and its location appears to rotate about the fixed Earth.

2.11. Action

There have been two fundamental descriptions of dynamics in physics, the equations of motion approach and the minimum action approach. They are equivalent, although the equations of motion, for example those of Newton, are rather more familiar. They yield second order differential equations which are typically solved by specifying initial positions and velocities. Several examples in classical mechanics appear in this chapter in fact.

The other approach is the "action principle". Simply stated the action, S, along the true path is an extremal, minimum, and the end points of the path are fixed. Alternative paths have an action larger

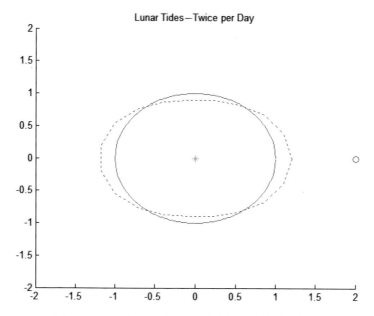

Figure 2.22: A frame from the movie provided by "Tides". The earth is red, the oceans are dashed blue and the Moon is a blue O.

than the path "chosen" by the system.

$$S = \int (T - V) dt \qquad (2.12)$$

Two examples are shown in the script "Lagrange_Act2", for motion in a uniform gravity field and for simple harmonic motion. The mass and harmonic frequencies are taken to be one for simplicity. The action in the two cases has the same kinetic energy, T, but differs in the potential energy, V:

$$S = \int \left(\frac{v^2}{2 - y^2}\right) dt, = \int \left(\frac{v^2}{2 - gy}\right) dt \qquad (2.13)$$

In the script the starting values at the twenty sample points are randomly chosen around the points of the exact solution. Then the MATLAB utility "fminsearch", is used to minimize the action. Therefore, each use of the script will yield a slightly different solution and the minimization, which is numerical, may not converge to strict

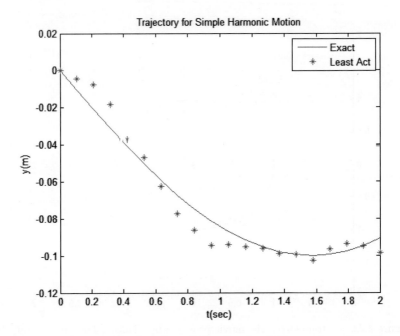

Figure 2.23: Exact and least action solution for simple harmonic motion. Twenty points in time are selected and the action is minimized with respect to them using "fminsearch".

limits. Typical results for both simple harmonic motion and for motion in a uniform gravity field appear in Figures 2.23 and 2.24 respectively. The minimization search is somewhat unstable, so that initial positions and velocities, number of sample points, and time span are defined by the script itself. The user may vary the parameters, but only by modifying the script. Aside from these caveats, the minimization of the action is made plausible by using this script. Using an enhanced number of points would presumably also smooth out the behavior, but at a cost of rather longer execution time for the search.

2.12. Rocket Drag

The simple rocket, as modeled, has no external forces acting on it. That model can be improved by taking the gravity field at the launch site into account. For a uniform field with acceleration g, an analytic

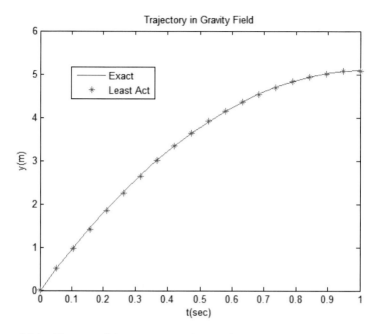

Figure 2.24: Exact and least action solution for motion in a uniform gravity field. Twenty points in time are selected and the action is minimized with respect to them using "fminsearch".

solution exists. An additional force arises from the drag due to the viscosity of the air through which the rocket passes. The height above the earth is y and v_o is the exhaust velocity. The acceleration drag depends on the rocket velocity, v, squared and the density of the local atmosphere, which falls off with altitude.

$$\frac{d^2y}{dt^2} = -v_o\frac{\left(\frac{dM}{dt}\right)}{M} - g - \frac{k\rho v^2}{M}$$
$$\rho = \rho_o e^{-\beta y} \tag{2.14}$$

The rocket mass is M and the mass burn rate, dM/dt, is assumed to be a constant. The gravitational acceleration is g and the air density is ρ, which falls off with altitude on an approximate scale $\beta = 1/9140\,\text{m}$. The drag constant, k, in MKS units is about 5000. This equation is solved numerically in the script "cm_rocket_drag" using the MATLAB utility "ode45". The advanced user could replace

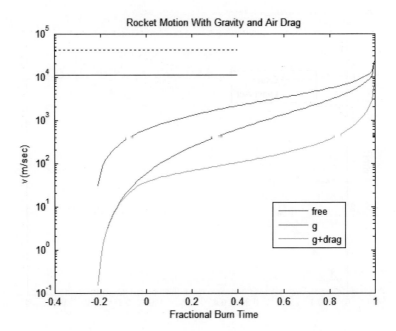

Figure 2.25: Velocity of a rocket which is free, in a constant gravity field, and subject to gravity and drag. The black horizontal lines are the escape velocity for the earth (solid) and sun (dashed).

g by the more correct gravitational force, $g \to g/(1+y/R)^2$, where R is the radius of the earth and study the resultant changes in the rocket trajectory.

The velocity of the rocket as a function of time is shown in Figure 2.25 for a particular case chosen by the user, near to the parameters of a Saturn V but with only a 1000 kg payload. In this example the payload attains escape velocity from the Earth. The free case goes fastest as expected. In the case with an added gravity field, the initial acceleration must become positive, leading to a delay in gaining altitude. Adding drag does nothing until the speed builds up as expected from the dependence on velocity squared. However, at later times when the velocity increases the air density falls off and the drag force drops.

The position of the rocket in the three cases is displayed in Figure 2.26. The free rocket is always highest at a given time. The

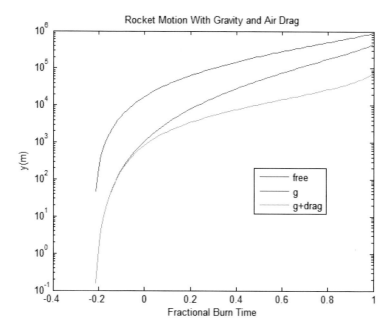

Figure 2.26: Height of a rocket which is free, in a constant gravity field, and subject to gravity and drag. The effect of viscosity drag is non-negligible.

effect of viscosity is not negligible and needs to be taken into account in any real application.

2.13. Two Stage Rocket

The basic rocket equation for exhaust velocity v_0 and initial and final masses of the rocket of m_i and m_f is found from applying momentum conservation to the system of the rocket itself and the ejected mass:

$$v_f = v_i - v_0 \ln\left(\frac{m_f}{m_i}\right) \qquad (2.15)$$

Gravity and drag are now ignored in the comparison of two and one stage rockets.

For a single stage rocket with initial mass m_0, payload mass m_p and ratio of plumbing mass to fuel mass rat the final mass is the payload plus the plumbing, so that the final velocity, with zero initial velocity is:

$$v_f = -v_0 \ln\left(\frac{(m_p + rat M_{fuel})}{m_0}\right) \qquad (2.16)$$

A two stage rocket can improve on the magnitude the final velocity by jettisoning the first stage dead weight of plumbing. Assuming the two stages have the same *rat* value, the same launch weight m_0 and payload mass m_p, the final velocity for a two stage rocket can be derived by repeated application of the basic rocket equation. It is:

$$m_{r1} = (m_0 - m_2)\left(\frac{1-1}{(1+rat)}\right)$$

$$m_{r2} = (m_2 - m_p)\left(\frac{1-1}{(1+rat)}\right) \qquad (2.17)$$

$$v_f = -\ln\left(\frac{(m_2 + m_{r1})}{m_0}\right) - \ln\left(\frac{(m_p + m_{r2})}{m_2}\right)$$

For the one stage the initial mass is m_0 and the final mass is the payload mass plus the plumbing for the total fuel load. The two stage case allows some of the plumbing to be shed with the first stage, which is assumed to be all dead weight plumbing when it is cast away. The plumbing masses are m_{r1} and m_{r2} for the two stages. The total second stage mass of payload plus plumbing plus fuel is m_2. The final two stage velocity is the velocity gain from the first stage plus the gain from the second stage.

The comparison is made in the script "Two_Stage". Dialogue from that script appears in Figure 2.27. The velocity comparison for the parameters given in Figure 2.28 and is plotted in Figure 2.29. A clear gain in final velocity is seen, which is why most space missions use multi-stage launch vehicles. The time unit T is the total fuel burning time assuming no plumbing, so that the actual times are less since dead weight is now explicitly assumed. The specific assumption is that the plumbing weight is 30% of the fuel weight.

There is an optimization that can be done for each particular set of parameters. For a 30% ratio of plumbing to fuel the results are shown in Figure 2.29. There is a weak optimal second stage weight

```
>> Two_Stage
   Rockets - 2 stage vs. 1 stage


Velocity, Satellite Low Circular Orbit (m/sec) = 7905.97
Escape Velocity - Earth (m/sec) = 11180.7
Escape Velocity - Solar System (m/sec) = 42092.7
Input the Rocket Mass (in 10^6 kg units) - Saturn = 4x10^6 kg: 4
Input the Payload Mass (in kg) - Saturn Escape Module = 24610 kg: 2000
Input the Exhaust Velocity (in m/sec) - Saturn = 2200 m/sec: 3000
Input Burn Rate (in kg/sec) - Saturn = 15000 kg/sec: 15000
Input Mass for Second Stage(in 10^6 kg units) : 1
Input the Plumbing to Fuel Mass Ratio for Both Stages : 0.3
```

Figure 2.27: Typical input parameters to compare a two stage and a one stage rocket.

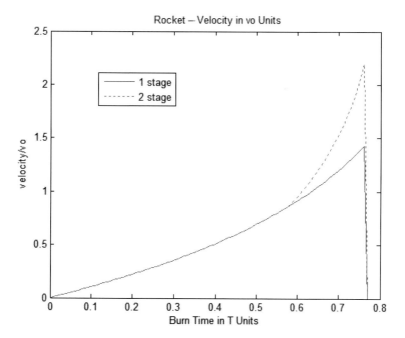

Figure 2.28: Comparison of the final velocities attained by a one and a two stage rocket.

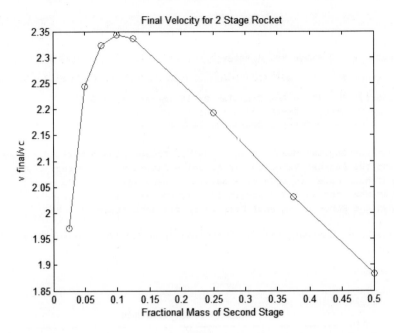

Figure 2.29: Final velocity as a function of the fractional mass taken by the complete second stage.

to total launch weight for these specific parameters which occurs at a second stage fractional weight of approximately 10%.

2.14. Table Top

This exercise consists of exploring the case of a mass m_1 on a horizontal plane surface connected to a mass m_2 by a string of length L hanging below the plane. There is a competition between the gravitational pull on the second mass and the centrifugal force due to the conserved angular momentum of the first mass point. There are well behaved solutions for moderate values of the angular momentum L. For too small values the lower mass pulls the upper mass down below the table. For too high values of L the upper mass pulls the lower mass up above the table.

The mass points obey the equations of motion, Eq. (2.18), which are solved numerically in the script "Table_Top" using the MATLAB tool "ode45". The user is encouraged to find the L value where the lower mass is just pulled up to z of zero and the other L value where the upper mass point approaches r of zero. The upper mass point has coordinates r and φ, while the lower mass point is at a distance z below the table.

$$\frac{d^2r}{d^2t} = \frac{\left(\frac{-gm_2 + L^2 m_1}{r^3}\right)}{(m_1 + m_2)}$$
$$\frac{d\phi}{dt} = \frac{L}{r^2}, \quad \ell - z = r$$
(2.18)

A plot of the motion of the two mass points is supplied as a movie and also as a three-dimensional plot using "plot3". As seen in Figure 2.30, the blue lines indicate the positions of the two mass points.

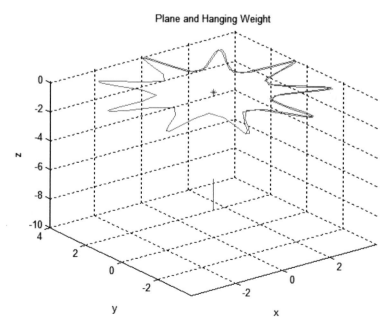

Figure 2.30: Plot of the positions of the two mass points. The upper mass point moves in the plane at $z = 0$, while the lower mass point moves up and down along the vertical axis z.

2.15. Kepler Eccentric

The simplest gravitational system to study is the motion of a light "test particle" in the field of a very heavy body. In that case, the motion of the heavy body can be neglected and the problem is solved when the motion of the light planetary body is specified. Since the force is central, motion of the planet is in a plane and angular momentum is conserved, leaving only two variables, the planetary radius and polar angle to be specified. It is fortunate that the solar system well approximates this effective one body planetary motion, to that the resulting simple Keplerian laws can be extracted from the motion of several, almost independent, planets.

The orbits for an inverse square law are conic sections; circles, ellipses, parabolas and hyperbolae. The character of the orbit depends on the eccentricity, e, which depends on the angular momentum of the planet. The circular orbits, with radius r_c, and radial velocity equal to zero are simplest.

$$r = r_c \frac{(1+e)}{(1+e\cos\theta)} \tag{2.19}$$

The orbits for several values of e are shown in Figure 2.31. The orbit at $e = 0$ is a circle, which evolves with increasing e into an ellipse and then a hyperbola. The script used is called "eccentric_Kepler". The printout of the script for a case where e is greater than one appears in Figure 2.32 which gives the asymptotic angle.

The boundary between bound and unbound orbits occurs at $e = 1$ when the degenerate ellipse opens up into a parabola. Note that, looking at Eq. (2.19), the radius r becomes large when $\cos\theta = 1/e$, which has a real angle as solution only if e is greater than one. The angles in question are the asymptotes. The velocities in the bound case are $v_y = v_c(1 + e\cos\theta)$ and $v_x = v_c e \sin\theta$ both scaled to the circular velocity

2.16. Non-Central Force

The solutions for a specific non-central force are explored in the script "Non_Central_Force". There is an additional harmonic force with

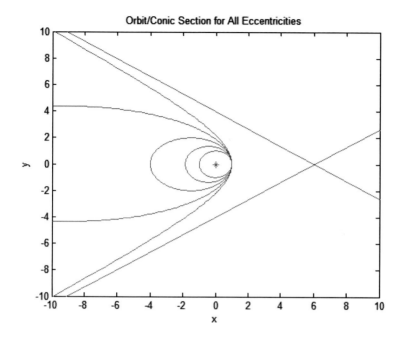

Figure 2.31: Orbits with eccentricity 0, 0.3, 0.6, 0.9 and 1.01, taking $r_c = 1$.

```
>> eccentric_Kepler
   Program to compute conic sections with different eccentricity


   Keplerian Orbits - defined by Eccentricity
   r = ro*(1+e)/(1+e*cos(theta))
   Enter Eccentricity, (0,2): 1.2
   e > 1 - Aysmptotic Angle = 33.5573
```

Figure 2.32: Printout from the script "eccentric_Kepler" for an unbound trajectory with $e = 1.2$.

amplitude b and harmonic frequency ω which is shown in Eq. (2.20).

$$\begin{aligned}\frac{d^2 r}{d^2 t} &= \left(\frac{d\theta}{dt}\right)^2 r - \frac{1}{r^2} \\ \frac{d}{dt}\left(\frac{r^2 \, d\theta}{dt}\right) &= b \sin(\omega t)\end{aligned} \qquad (2.20)$$

The gravitational force appears with unit strength. In the script, a symbolic solution is first tried but it is not successful. Then a numerical solution using "ode45" is adopted. The user dialogue allows for a choice of the strength of the added force and the angular frequency. The major axis of a Keplerian orbit and the angular velocity are then chosen, which defines an elliptical orbit in the absence of the additional force.

The perturbing force distorts the ellipse, elongates it and causes it to be non-reentrant. The angular momentum is no longer conserved, Eq. (2.20), so that the problem is intrinsically two-dimensional. For small amplitudes the effect is small. For high frequencies with respect to the main orbital motion the effect is also washed out since the averaged force approaches zero. Typically, for a high enough frequency and a small enough amplitude, the force only causes the ellipse to not close on itself, as illustrated in Figure 2.33. At larger

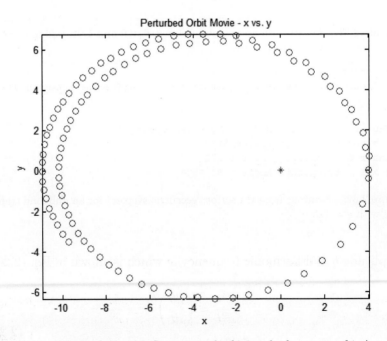

Figure 2.33: Movie of the orbit for a strength of 0.3 and a frequency of 1. A major axis of 4 is assumed for the Keplerian ellipse.

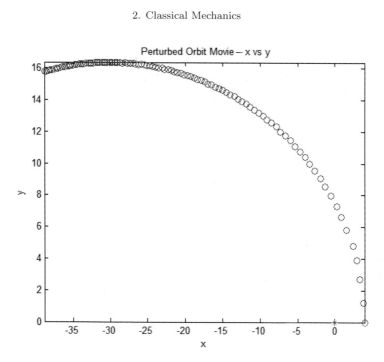

Figure 2.34: Movie of the orbit for a strength of 0.6 and a frequency of 1. A major axis of 4 is assumed as the starting point of a force free ellipse. The actual orbit is nearly unbound due to the additional force.

amplitudes the effect of the added force becomes more pathological, as the user can easily explore.

A stronger force with a low frequency will distort the orbit even more. A low frequency means that the sign of the force does not change rapidly during the orbit and thus the force pushes the orbit to large radii. The strength is then sufficient to make the bound orbit into a free orbit. A movie for this situation is shown in Figure 2.34.

2.17. Euler Angles

The Euler angles were created to relate the position of a rigid body in its own rest frame (body frame) to the position in a frame where the body is in motion (space frame). Indeed, the transformation has already been invoked in the multiple scattering script which was used in the exercise in the study of a muon beam, where the multiple scattering with respect to the muon direction was transformed to the angles in the laboratory (space) frame through which the muon propagated. In that script a special case of the general transformation, only two angles needed, was used.

The general transformation first rotates about the z axis, followed by a rotation about the x axis and then a second rotation about the z axis. The cosine and sin, c and s, of the three angles are indicated in Eq. (2.21) below. In the figure, all angles are 35°.

$$\rightarrow \begin{pmatrix} c & s & 0 \\ -s & c & 0 \\ 0 & 0 & 1 \end{pmatrix} \rightarrow \begin{pmatrix} 1 & 0 & 0 \\ 0 & c1 & s1 \\ 0 & -s1 & c1 \end{pmatrix} \rightarrow \begin{pmatrix} c2 & s2 & 0 \\ -s2 & c2 & 0 \\ 0 & 0 & 1 \end{pmatrix} \quad (2.21)$$

A visualization of the effect of these rotations is provided by the script "Euler_Angles" where the script output plot appears in Figure 2.35. The locations on a circle and the x and z axes are indicated by blue, solid → red, dash dot → green, dashed → black dotted for the

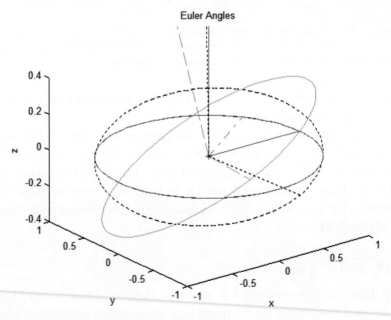

Figure 2.35: Euler angle rotations from body → space frames showing the x and z axes and a circle in the (x, y) plane for the three successive rotations, blue is initial, red is after the first rotation, followed by green and then black. The orientation is indicated by the x and z axes and a circle in the (x, y) plane in each case.

initial axes and then the results of the three successive rotations. The first rotation is in the (x, y) plane, while the second is about the transformed x axis. The final third rotation is about the transformed z axis. The specification of the position of a rigid body requires three angles in Euclidian three-dimensional space, assuming the C.M. velocity is zero.

The Euler angles will be illustrated by the study of top motion. However, they are perfectly general and can be applied to any situation describing rigid body motion.

2.18. Forceless Top

A simple example of rigid body motion is that of a top spinning without external forces acting on it. The shape of the top is defined by the moment of inertia about the rotation axis, I_z, and the moments about the other principle axes, I_1, assuming that the top is rotationally symmetric. The top axis of rotation precesses in this case with a frequency $\omega_{pre} = [(I_z - I_1)/I_1]\omega\cos\theta$ where ω is the top rotational frequency and the top axis is inclined by the angle θ with respect to the vertical. The precession frequency depends on the top being non-spherical.

An illustration of the motion is given in the script "Top_Forceless". A specific output plot is shown in Figure 2.36. The user can change the shape of the top and its inclination angle and gain some insight on the subsequent motion. In this case the top rotational axis simply precesses with circular frequency ω_{pre} about the z axis at a constant angle of inclination.

2.19. Top Motion

An illuminating use of the Euler angles is the study of top motion. The top is acted on by a torque due to gravity which then changes the angular momentum of the top. The top is assumed to be rotationally symmetric which leads to a constant of the motion, ω_s. There remain five variables to solve, for the three angles specifying the top and their derivatives, less the constant of motion. The torque is defined by the

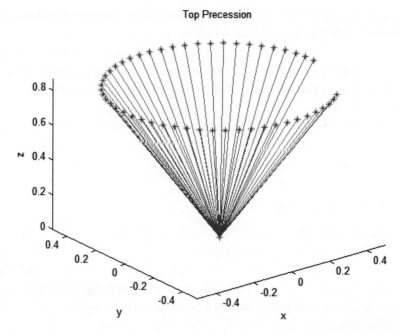

Figure 2.36: Precession of a top not acted on by any forces. The axis of rotation is indicated by the red lines as a function of time in the movie provided. The top has a shape with $I_z = 1.5$ and $I_1 = 1$ and the inclination angle is 30 degrees.

mass of the top, m, and the height of the C.M. of the top above the fixed point of rotation, h.

$$\omega_s = \frac{3}{2} + \frac{\sqrt{4I_1 mgh\cos\theta}}{I_z}, \quad B = mgh \qquad (2.22)$$

The rotation axis is z with moment of inertia in the body frame defined to be I_z, while the other axis is defined to be, as before, I_1. In general the top rotational axis precesses, as in the forceless case, although the precession rate is now not a constant. It also nutates, with a complex path swept out in the polar angle variable. This motion has the top initially dropping under gravity and then rising in a periodic fashion. In fact, with extreme values of the user supplied parameter, the top will "tumble" and then right itself later.

The equations to be solved in the space frame are:

$$\frac{d^2\phi}{dt^2} = \frac{\left[\left(\frac{I_z\omega_s - 2I_1\,d\phi}{dt\cos\theta}\right)\frac{d\theta}{dt}\right]}{I_1 \sin\theta}$$

$$\frac{d^2\theta}{dt^2} = \frac{\left[mgh - \left(\frac{I_z\omega_s - I_1\,d\phi}{dt\cos\theta}\right)\frac{d\theta}{dt}\right]}{I_1} \quad (2.23)$$

$$\frac{d\psi}{dt} = \frac{\omega_s - d\phi}{dt\cos\theta}$$

Since the top is not translating there are three variables specifying the top; the polar and azimuthal angles of the top axis in the space frame and the angle ψ defined by the spin of the top itself about its axis of rotation.

This set of equations is solved numerically in the script "Top_Euler" using the MATLAB utility "ode45". The initial azimuthal angle φ is zero, while the initial polar angle θ, which defines the inclination of the top, is defined by the user. The initial value of the time rate of change of the azimuthal angle sets the behavior of the nutation, and it is also a user supplied input. The spin is defined by the angle ψ. In general there are five functions to determine. The dialogue is shown in Figure 2.37.

```
>> Top_Motion
   Top Motion using explicit solution, constants of the motion


Top motion depends on spin, Top moments of I, mass
and initial posiition and velocity (4 of)
Special Case of "drop", theta = phi = phidot = 0, only thetao
Enter 2*m*g*h/I1 -Torque, units 1/t^2 (200) : 200
Enter Top Shape Iz/I1, I1=1 (1.2) : 1.4
Enter Top Spin (rad/sec) (30) : 30
Enter Initial Top Polar Angle (deg) (20) : 30
```

Figure 2.37: User definition of the motion of a top. The top mass and shape are user defined as are the top spin and angle of inclination from the vertical.

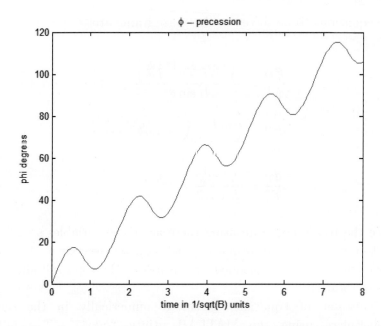

Figure 2.38: Precession of a top with gravitational torque acting on it. The structure in time is no longer simple as it was in the forceless case.

A typical result for the precession in azimuth is shown in Figure 2.38. The rate is not constant, as in the forceless case, but is modulated by the nutational motion in polar angle.

In Figures 2.39 and 2.40 the nutational pattern for the polar angle is shown for two values of the initial time rate of change of the azimuthal angle. In Figure 2.41 the last frame of a movie of the motion of the top axis is shown, illustrating the precession and nutation.

If the time rate change of initial azimuthal angle is set to zero, the top merely drops initially in polar angle until if rises and executes simple harmonic motions. There is another special case, where the top is initially vertical, the "sleeping top", where the top then simply spins about the z axis as in the forceless case because the torque is then zero. The user has several parameters to explore in this exercise. The motion is complex and the user is highly encouraged to vary the top defining parameters and see what the subsequent motion is. The top may "sleep", or simply precess, or fall below $z = 0$.

2. Classical Mechanics

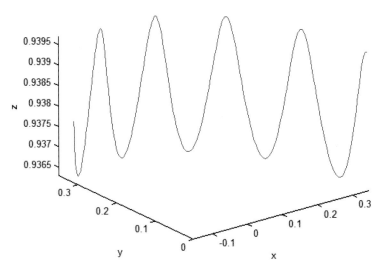

Figure 2.39: Nutation in the case of a small positive value for the time rate change of the azimuthal angle.

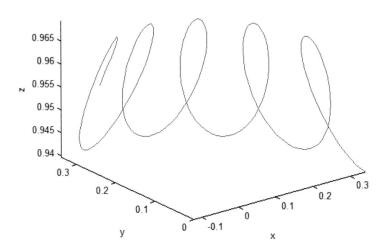

Figure 2.40: Nutation in the case of a larger positive value for the time rate change of the azimuthal angle.

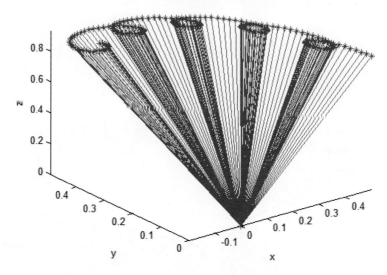

Figure 2.41: Motion of the axis of the top for the case of a 30-degree initial angle and initial azimuthal time rate of change as in Figure 2.39.

Chapter 3

Electromagnetism

"One day sir, you may tax it."
— **Michael Faraday**'s reply to William Gladstone, then British Chancellor of the Exchequer, when asked of the practical value of electricity

"Magnetism, as you recall from physics class, is a powerful force that causes certain items to be attracted to refrigerators."
— **Dave Barry**, Pulitzer Prize-winning American author and humor columnist

3.1. Electron e/m

The electron was the first fundamental particle of matter to be discovered. Since electrons are bound in atoms only by a few electron-volts, they can be relatively easily liberated from atoms. The apparatus is shown schematically in the output of the script "Electron_e_m". Electrons are boiled off the cathode of a vacuum tube. They are then accelerated in an electric field indicated in Figure 3.1 below as the black boundary lines. The acceleration due to the vertical electric field causes the path to curve, shown as a red line. Applying a magnetic field using the Helmholtz coil indicated as four blue o, the path can be restored to be a straight line as confirmed using a phosphor screen at the rightmost end of the vacuum tube.

The user can derive the condition for no net force with a known electric, E, field along y and a magnetic, B, field along z. The measured length is L and the measured displacement in the case where the magnetic field is zero is y. The initial x velocity is v set by applying a voltage to first accelerate the electrons before they enter the region with the fields. Therefore the charge to mass ratio of the

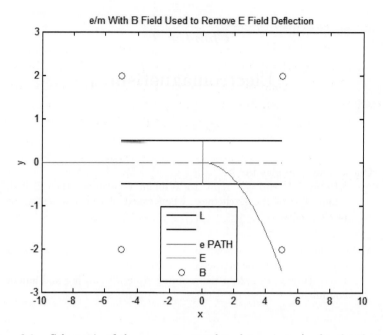

Figure 3.1: Schematic of the apparatus used to determine e/m for the electron. The e without the B field is deflected by the vertical E field by y. When the B field is adjusted so that there is no deflection, the e/m ratio is determined.

electron can be measured using this technique. The distance between source and screen is L and t is the transit time.

$$\vec{F}_B = e(\vec{v} x \vec{B}), \quad \vec{F}_E = e\vec{E}$$
$$v_x = \frac{E}{B}$$
$$y = \frac{\left(\frac{eE}{m}\right)t^2}{2}, \quad t \sim \frac{L}{v_x} \tag{3.1}$$
$$\frac{e}{m} \sim \frac{2yE}{(L^2 B^2)}$$

In Eq. (3.1) all quantities are measured on the right-hand side of the expression for e/m. In this fashion J. J. Thomson was able to determine the charge to mass ratio of the electron.

3.2. Potential and Fields

Conservative forces like gravity are derivable from potentials. The fields are the gradient of the potentials, $\vec{E} = -\nabla V$. The script "Potential_Field" asks for a symbolic input potential in two dimensions, plots it and then finds the x and y fields due to that potential. This is done symbolically using the MATLAB utility "diff" to find the fields with a defined potential. The script is therefore useful in visualizing the shapes for any potential that can be expressed analytically. The potential, $V = \tanh(xy)$ is plotted in Figure 3.2, and the x field for that potential appears in Figure 3.3. The user is encouraged to explore other potential shapes which might interest him/her.

3.3. Electrostatics — Plates

The problem of a charge located at $x = 0$ and $y = d$ between two grounded plates at $y = 0$ and $y = a$ is considered in the script

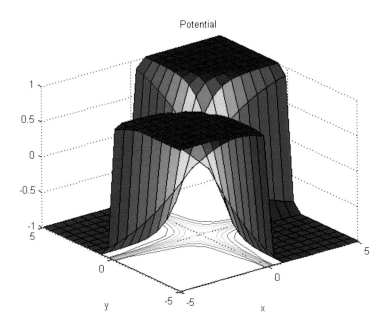

Figure 3.2: Potential and contours for $V = \tanh(xy)$.

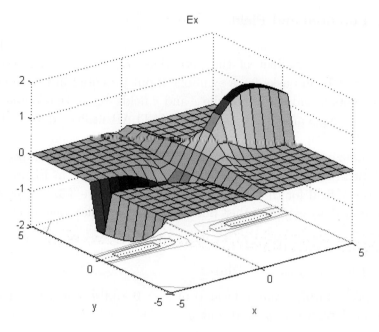

Figure 3.3: Field in x for the potential $V = \tanh(xy)$.

"EM_Plates". The potential is assumed to factorize into a function of x times a function of y. The solution is a series of terms which vanish at the conducting plates. The solution for $x > 0$ has a sign change in the argument of the exponential compared to the solution for $x < 0$ which is shown in Eq. (3.2).

$$\Phi(x<0) = \sum_n 4q \sin\left(\frac{n\pi d}{a}\right) \sin\left(\frac{n\pi y}{a}\right) \frac{e^{n\pi x/a}}{n\pi} \qquad (3.2)$$

The user chooses a value of d/a and the number of terms in the series. The potential for the specific choice d/a of 0.8 and 20 terms appears in Figure 3.4 for the potential. That function peaks at the location of the point charge and vanishes on the plate boundaries as expected. The electric field in the y direction is evaluated using the MATLAB numerical utility "gradient" and is shown in Figure 3.5.

The user should vary the number of terms to get a feeling of how quickly the series converges to a smooth behavior as a function of x

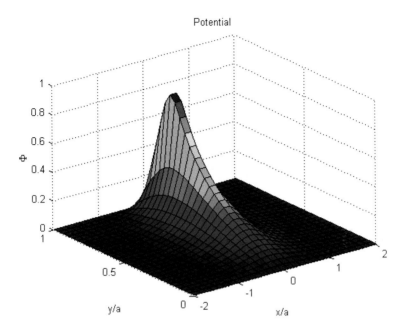

Figure 3.4: Potential for the case where $d/a = 0.8$ for a series with 20 terms.

and y. The potential must vanish at the plate boundaries at y equal to zero and a and must fall of rapidly as x differs from the charge location at x of zero.

3.4. Electrostatics — Angled Plates

A different but similar geometry is explored in the script "EM_Angled". In this case the grounded plates intersect at the origin and are at an angle α with respect to one another. The point charge is located between the plates at an angle β at a radius r_o. The observation point is defined by the radius r and the angle θ. The potential is again solved as a series. However, the geometry is no longer Cartesian but polar so that the series is not a Cartesian Fourier series but a power series in radius.

$$\Phi(r < r_o) = \sum_n 4q \left(\frac{r}{r_o}\right)^{n\pi/\alpha} \sin\left(\frac{n\pi\beta}{\alpha}\right) \sin\frac{\left(\frac{n\pi\theta}{\alpha}\right)}{n} \qquad (3.3)$$

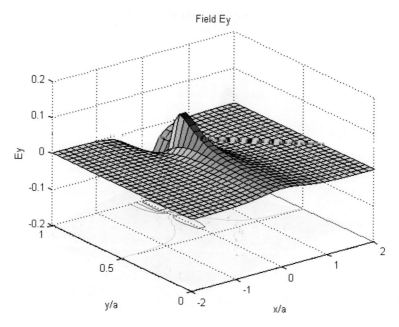

Figure 3.5: Electric field in the y direction. There is a normal field which reflects the induced charge on the upper plate.

The solution for r greater than r_o inverts the power to be (r_o/r). This ensures that the potential vanishes at the origin and falls off at large r values, as is physically reasonable. The potential is displayed as a function of x and y after the user defined values of α and β and the number of terms of the series are given. The result for $r_o = 1$ with a thirty degree angle of the plates and a ten degree angle for the charge are shown in Figure 3.6.

The contours indicate that the plates are equipotentials which therefore satisfy the boundary conditions. There is a general falloff with distance from the point charge. Another example is shown in Figure 3.7. In this case α is sixty degrees and β is forty degrees. The change in the equal potential contours is very evident with respect to the previous example. The vertical field near the lower plate, $E_y(x,0)$, is shown in Figure 3.8. It was derived using the potential and the MATLAB utility "gradient". The x electric field at the lower plate is small, zero within the numerical accuracy as is required by

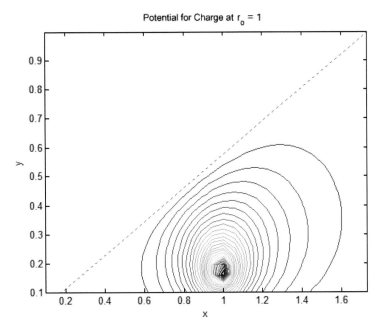

Figure 3.6: Potential contours for the case where $\alpha = 30$ degrees and $\beta = 10$ degrees. The plates are located at $y = 0$ and the second plate is indicated by the red dashed line.

the boundary conditions. The user is encouraged to try several values of the two angles.

3.5. Electrostatic Helmholtz

The Helmholtz coil consists of two magnetic coils such that the magnetic field between them is rather uniform spatially. The electrostatic analogue is two rings, radius a, of charge plus and minus Q with an axial distance L between them. The dialogue for the script "Electric_Helmholtz" for a radius a of one appears in Figure 3.9. The field is axial by symmetry.

For the specific relationship between the separation L and the radius a, $L = \sqrt{6} * a$, the first and second derivatives both vanish in the center of the two loops. The solution is found using "factor" on the second derivative evaluated at z of zero in the script. The field and the derivatives for the special relationship are shown in

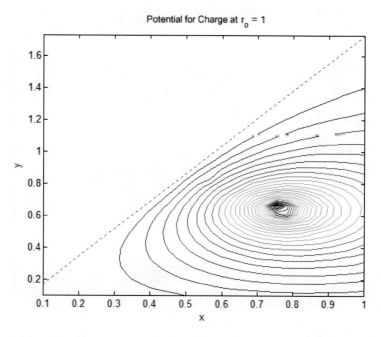

Figure 3.7: Potential contours for the case where $\alpha = 60$ degrees and $\beta = 30$ degrees. The first plate is located at $y = 0$ and the second plate is indicated by the red dashed line.

Figure 3.10. The field is quite constant over the full L range. Both the first and second derivatives vanish in the center of the rings which implies a rather constant field along the axis.

3.6. Electric Quadrupoles

The electric field has various multipole components which have characteristic angular and radial dependencies. The field due to an isolated charge goes as $1/r^2$ and is isotropic. For a dipole the field goes as $1/r^3$ and has a $\cos\theta$ distribution for a dipole moment oriented along the z axis. For a quadrupole field the radial dependence is $1/r^4$ with an angular dependence, for Qzz the sole quadrupole moment component, of $3\cos^2\theta - 1$.

The script "EM_Quadrupole" treats the quadrupole field symbolically, using the utility "diff" to find the fields. The basic charge distribution can be thought of as a superposition of two dipoles

3. Electromagnetism   67

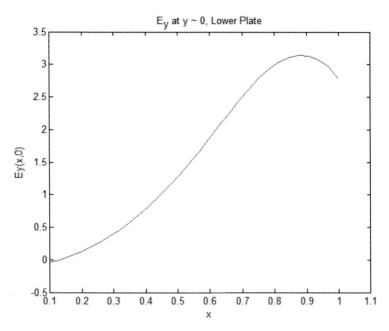

Figure 3.8: Field normal to lower plate for the potential shown in Figure 3.7. The field Ey for $y = 0$ vanishes at the origin and peaks around the x location of the charge between the plates.

```
>> Electric_Helmholtz
   2 charged rings, +- Q, electrostatic analogue of Helmholtz coil


Two Charged, + - , Rings, z = 0 and z = L
Field Ez and First and Second Derivatives

          z                  L - z
     -----------    +    ------------------
        2    3/2            2      3/2
     (z  + 1)           ((L - z)  + 1)
```

Figure 3.9: Printout of the script "Electric_Helmholtz". The first and second derivatives of the field are prints by the script but are not shown here. The MATLAB symbolic utility "diff" is used to derive the fields and derivatives so as to explore the uniformity of the field.

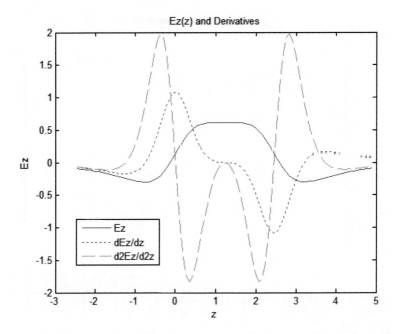

Figure 3.10: Plots of the field and its derivatives for the special case of a uniform field, where $L = \sqrt{6} * a = 2.45a$ where a is $=1$ in this plot.

oriented in opposite directions. There are single charges at $+z$ and $-z$ and an opposite doubles charge at z of zero. The printout made by the script appears in Figure 3.11. A plot of the radial dependence of a quadrupole, both potential and field appears in Figure 3.12. The inverse fourth power of the field appears even rather close to the quadrupole structure.

A plot of the spatial shape of a quadrupole with only a z, z component of the quadrupole moment is plotted in Figure 3.13. The behavior clearly has two lobes, while a monopole shape has no poles and a dipole shape has only one pole.

3.7. Cylinder Boundaries

Consideration of the electrostatic problem of a long cylinder, radius a, where the positive y half cylinder is at potential V_o while the bottom is at $-V_o$ is given in the script "Laplace_Cyl". The Laplace equation is appropriate as there are no sources, only specified boundary values.

3. Electromagnetism

```
>> EM_Quadrupole
   Program to look at Quadrupole electric fields

Charge has a field ~ q and 1/r^2
Dipole has a field ~ qd and 1/r^3
Quadrupole has a field ~ qd^2 and 1/r^4
Charge +1 at z = a, -2 at z = 0 and +1 at z = -a

V =

1/(a + r) - 2/r - 1/(a - r)


E =

2/r^2 - 1/(a + r)^2 - 1/(a - r)^2
```

Figure 3.11: Symbolic calculation of the potential and the radial electric field for a simple quadrupole arrangement of charges. An expansion of the field in small values of a/r reveals the inverse fourth power of the radius as the leading term in the field.

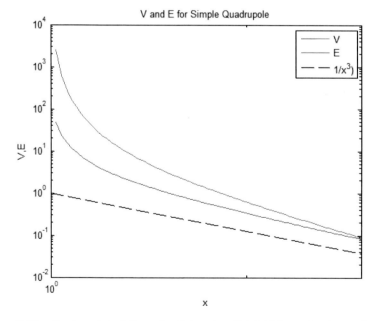

Figure 3.12: Loglog plot of the potential and radial field for a simple quadrupole layout. The x axis is $(1, 3)$ in units of a. The line with $1/r^3$ dependence is to compare to the potential r dependence. The power law behavior sets in quite rapidly, with r/a less than 2.

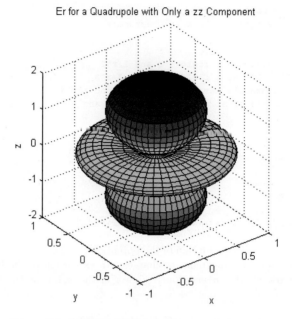

Figure 3.13: The solid shape derived for the spatial dependence of the radial field of a quadrupole with a moment possessing only a z, z component.

Both the region interior and exterior to the cylinder is solved using a series method. The small radius solution vanishes at the origin, while the large radius solution falls off rapidly with radius. The potential, in V_o units, with indices that are odd is:

$$A_i = \frac{4}{(i\pi)}$$
$$V_{in} = \sum_i A_i \sin(i\phi)^* r^i \quad (3.4)$$
$$V_{out} = \sum_i \frac{A_i \sin(i\phi)}{r^i}$$

The script assumes that the radius, a, is one as is the potential magnitude at that radius. The user chooses the number of terms in the series, which allows a visual inspection of the convergence of the series. The electric fields are derived numerically using the MATLAB

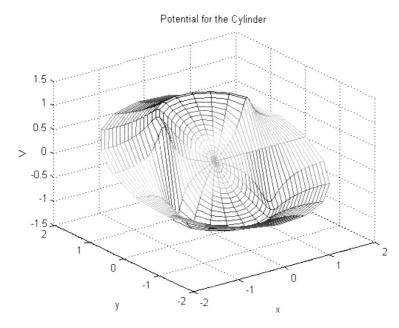

Figure 3.14: Potential both interior and exterior to a cylinder of radius one and potential one. The rise in V at interior r and the falloff at large r are evident.

"gradient" tool. The potential for a series with twenty odd terms is shown in in Figure 3.14, while the y component of the electric field appears in Figure 3.15. With twenty terms the representation on the radial boundaries appears to be quite good in that the constancy of the potential is manifest.

3.8. Laplace Fourier

Boundary problems with the Laplace equation can be solved for Cartesian boundary conditions by variants of the Fourier series technique. The related problem for cylindrical boundary conditions appears in Section 3.7. For the case of a conducting box of length $2a$ in x and $2b$ in y, with the left and right sides at $x = -a$ and $x = a$ respectively held at voltages V_L and V_R and top and bottom grounded the solution for the interior potential is a series with only

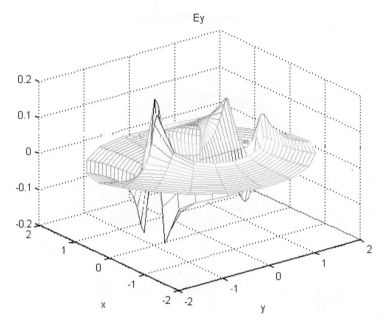

Figure 3.15: The field derived from the series from the potential of Figure 3.14. The behavior is complex at the boundary along $y = 0$.

odd terms

$$V(x,y) = \sum_i [A_i e^{-i\pi x/a} + B_i e^{i\pi x/a}]\sin\left(\frac{i\pi y}{b}\right)$$

$$A_i = \left(\frac{4}{i\pi}\right)\frac{[V_L - V_R e^{-i\pi(a/b)}]}{c}$$

$$B_i = \left(\frac{4}{i\pi}\right)e^{-i\pi(a/b)}\frac{[V_R - V_L e^{-i\pi(a/b)}]}{c}$$

$$c = 1 - e^{-i2\pi(a/b)}$$

(3.5)

The script "Laplace_Series" covers this situation. The user supplies the parameters, a, b and the left and right voltages. In addition the number of terms in the series is asked for so that the user can see how rapidly the convergence of the series is. A mesh plot for the case of a equal to two, b equal to one, left voltage of

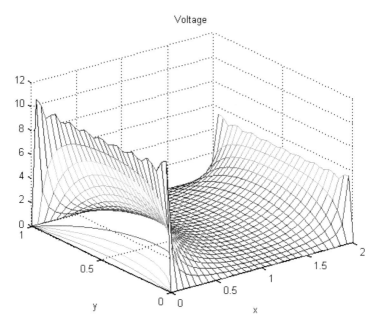

Figure 3.16: Voltage inside the box with sides at $y = 0, 1$ grounded and sides at $x = 0, 2$ held at 10 volts and 5 volts, respectively.

10 and right voltage of 5 with twenty terms in the series is shown in Figure 3.16. The left and right constant voltages on the boundaries are clearly seen, as well as the voltage fall from left to right.

The electric field in the x direction for this specific case is shown in Figure 3.17. It is large in the four corners of the box where the potential changes rapidly as expected.

3.9. Rotating Charged Sphere

A rotating charged sphere, radius a, will have an associated current which will produce a magnetic field, both inside and outside the sphere. The solution for this problem appears in Eq. (3.6). The axis of rotation is z. The solution can be thought of as a superposition of many current loops, found by integration. The units are indicated in Eq. (3.6) which factor out the rotational frequency and charge as

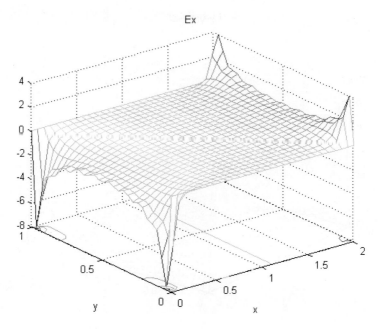

Figure 3.17: E_x inside the box with sides at $y = 0, 1$ grounded and sides at $x = 0, 2$ held at 10 volts and 5 volts, respectively.

well as numerical factors

$$\frac{(\mu_o Q \omega)}{10\pi} = 1$$

$$B_{r,out} = \frac{a^2 \cos\theta}{r^3} \tag{3.6}$$

$$B_{r,in} = \frac{\cos\theta(5a^2 - 3r^2)}{2a^3}$$

The current depends on the charge, Q, and the rotational frequency ω. The coordinates r and θ are spherical polar, and the problem is azimuthally symmetric. The magnetic field is shown in Figure 3.18. The field has a dipole like angular and radial dependence for radii greater than a, taken to be equal to one in the mesh plots. The interior field at the origin is $\cos\theta(5/2a)$, while it falls to $\cos\theta/a$ at the surface of the sphere, matching the exterior solution for the field.

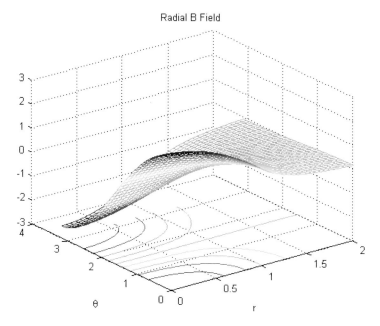

Figure 3.18: Mesh plot for the interior and exterior radial magnetic field of a rotating charged sphere.

The field in the polar angle direction is defined in Eq. (3.7) and is shown in Figure 3.19.

$$B_{\theta,out} = \frac{a^2 \sin\theta}{2r^3}$$
$$B_{\theta,in} = \frac{\sin\theta(-5a^2 + 6r^2)}{2a^3}$$
(3.7)

The angular field peaks at ninety degrees to the axis of rotation which is typical of a dipole field. At larger radii the field falls as $1/r^3$. The interior field is largely negative and rises as r^2, matching the exterior solution at radius of a.

3.10. Dielectric Image Charge

Dielectrics have polarization charges which change the boundary conditions which are appropriate to vacuum and to perfect conductors. There is a displacement vector field D which is the dielectric constant

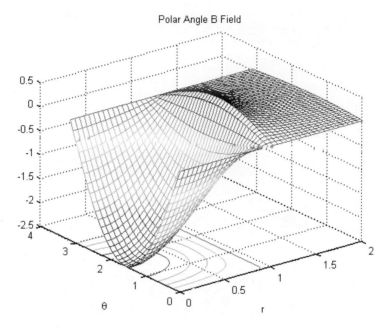

Figure 3.19: Mesh plot for the interior and exterior angular magnetic field of a rotating charged sphere.

times the electric field. The tangential electric field is continuous across a boundary between differing dielectric objects as is the potential, while the normal electric field is not. The normal D field is continuous.

The simple case of a point charge, q, located at a distance a above a dielectric medium filling the half space defined by z less than zero. The dielectric constant for that region is K. At the boundary E_x is continuous as is the potential, V, but E_z is not. The problem is explored in the script "Dielectric_Image". The method of image charges is used to satisfy the boundary conditions. The symbolic MATLAB utility "diff" is used to derive the fields from the potential and some of the results which are printed out appear in Figure 3.20.

For z greater than zero the potential has a term due to the origional charge q located at z of a, and an image charge $-q(K-1)/(K+1)$ located at z of $-a$. The upper plane is assumed to be vacuum.

3. Electromagnetism

```
Dielectric Plane with Dielectric Constant K, z < 0
Charge q at z = a in vacuum, K = 1
Potentials for z > 0 and z < 0

             1                           K - 1
   ------------------- - -------------------------------
        2      2 1/2              2          2 1/2
   (x  + (a - z) )         (K + 1) ((a + z)  + x )

             1                           K - 1
   ------------------- - -------------------------------
        2      2 1/2          2      2 1/2
   (x  + (a - z) )        (x  + (a - z) )       (K + 1)
Ex for z > 0 and z < 0

             x (K - 1)                         x
   ------------------------------- - -------------------
            2          2 3/2             2      2 3/2
   (K + 1) ((a + z)  + x )           (x  + (a - z) )

             x (K - 1)                         x
   ------------------------------- - -------------------
        2      2 3/2                      2      2 3/2
   (x  + (a - z) )       (K + 1)      (x  + (a - z) )
```

Figure 3.20: Printout of "Dielectric_Image" which gives the potentials appropriate for $z > 0$ and for $z < 0$ and E_x for the two regions. The field E_z is also printed out by the script but is not shown here.

In the lower half plane there is the potential due to charge q at z of a and an image term as before, but now also located at z of a. This potential is shown in Figure 3.21, in the specific case of a of one and K of four.

The fields E_x and E_y are evaluated after being derived symbolically from the potential. Both fields on the z axis at x equal to zero are shown in Figure 3.22. Clearly the tangential field is continuous, while the normal field is discontinuous, though the potential itself is continuous. The net result is to decrease the fields in the dielectric.

The user is encouraged to try some limiting cases and see if intuition is verified. For K equal to 1, there should be no boundary. For finite K, E_z will be discontinuous at the boundary while E_x is

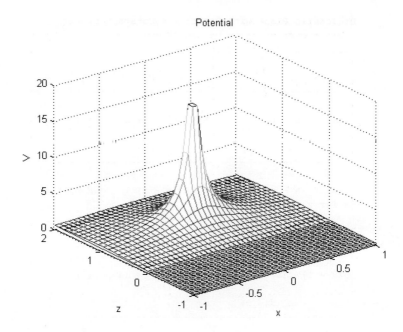

Figure 3.21: Potential as a function of x and z for a charge at $z = 1$ above a half plane with dielectric constant of 4.

continuous. For a very large K, a conductor is approached as a limit, so that $Ex(0)$ approaches zero in the upper half plane.

3.11. Dielectrics

The problem of a conductor surrounded by a dielectric cylinder immersed in a uniform electric field, E, is explored using the script "Dielectric". The user chooses the two radii, a inner and b outer, and the dielectric constant, K, of the cylinder. The solutions for the potentials outside and inside the dielectric cylinder are:

$$V_{out} = \left(\frac{Er + A}{r}\right)\cos\theta$$
$$V_{in} = \left(\frac{Br + C}{r}\right)\cos\theta$$
(3.8)

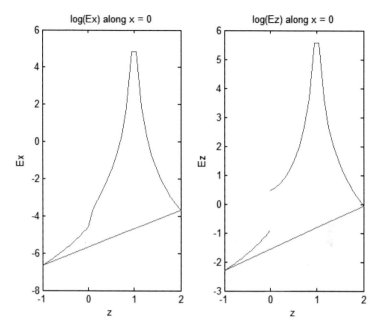

Figure 3.22: Plot of the z dependence of the tangential (left) and normal (right) field at $x = 0$. The log of the field is plotted to bring out the behavior at the boundary at $z = 0$.

The constants A, B and C are found by matching the boundary conditions at the two radii, and the results can be found in the script, "Dielectric". The exterior potential approaches the applied field E_z since the term A/r falls off at large r. Inside the dielectric terms that go as $1/r$ and r both exist. For r less than a, the conductor, the fields are zero.

The potentials for a choice of one for radius, a, of the conductor (red), and two for the dielectric radius, b, (green) and a dielectric constant of five are shown in Figure 3.23. There is no field in the conductor as in true for all electrostatic shielding and the surface at radius a is an equipotential. From the potential plot it is clear that there are both normal and tangential fields at the interface at a radius of b.

Note that the potential lines are not perpendicular to the green boundary as they would be for a conductor and are at a radius of

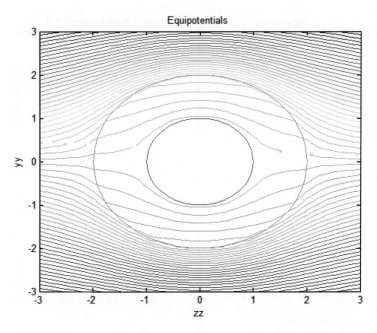

Figure 3.23: Equipotentials for a specific example of a dielectric cylinder surrounding a conductor and immersed in a field which is uniform in z at large distance.

a for the surface of the conductor. The user can vary the dielectric constant in order to see how the boundary conditions change the potential.

The electric fields derived from the potentials are plotted in Figure 3.24, which used the MATLAB plotting utility "quiver". The fields are not perpendicular to the outer, green boundary. They also have a large angular dependence. Inside the conductor the fields are zero to the level of numerical accuracy.

3.12. Dielectric Sphere

A dielectric sphere immersed in a uniform electric field, E_z does not result in a zero field inside the sphere as is the case for a conductor. For the case of a dielectric constant K the potentials inside the sphere and outside and the induced surface charge, σ, normalized to the

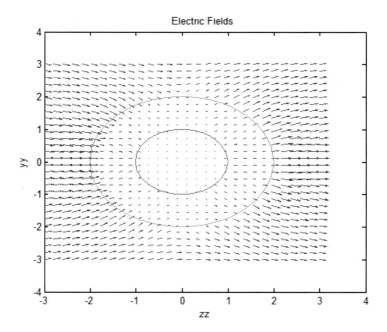

Figure 3.24: Electric fields for a specific example of a dielectric cylinder surrounding a conductor and immersed in a field which is uniform at large distance. The field inside the conductor is numerically zero.

external electric field are:

$$\frac{V_{in}}{E_o}\cos(\theta) = \frac{-3}{[(K+2)]r}$$

$$\frac{V_{out}}{E_o}\cos(\theta) = \frac{-\left[\frac{1-(K-1)}{(K+2)}\left(\frac{a}{r}\right)^3\right]}{r} \quad (3.9)$$

$$\frac{\sigma}{E_o}\cos(\theta) = \frac{\left(\frac{3}{4\pi}\right)(K-1)}{(K+2)}$$

At a radius of a the potentials match to a normalized value of $-3/(K+2)*a$ and the potential is continuous across the boundary. If the dielectric constant is equal to one, vacuum, then the induced charge is zero as expected. The script "Dielectric_Sphere" plots the solution for a user defined dielectric constant. The equipotentials for a constant K equal to five appear in Figure 3.25 and the derived

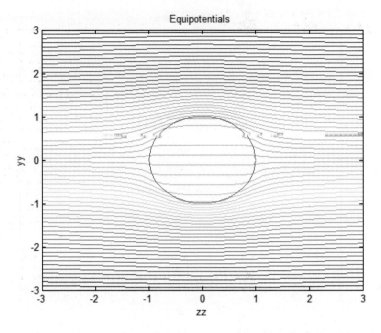

Figure 3.25: Equipotential contours for a dielectric sphere with constant K equal to five. There is no y gradient inside the sphere. On the boundary there are both normal and tangential fields.

electric fields are shown in Figure 3.26. Clearly, the field penetrates into the sphere. The user can view the depth of penetration while varying the dielectric constant. The special case of K equal to one is also instructive.

The electric fields of Figure 3.26 are again derived from the potential using the numerical MATLAB utility "quiver". The interior fields are horizontal, and the vertical field is essentially zero. The induced surface charge goes as the cosine of the angle as does the external electric field.

3.13. Induction

An electromotive force, EMF, is induced in a circuit when magnetic flux lines are time dependent, an effect first shown by Faraday. Consider the simple case of a moving charge, q, which creates both

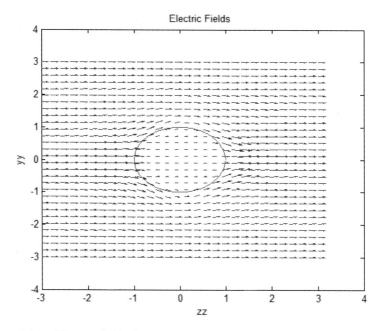

Figure 3.26: Electric fields for a dielectric sphere with constant K equal to five immersed in a uniform electric field along the z axis. The field inside the sphere is non-zero and largely along the z axis.

electric, E, and magnetic, B, fields

$$\vec{E} \sim \frac{q\vec{r}}{r^3}, \quad \vec{B} \sim \vec{v} x \vec{E}$$
$$EMF \sim qv^2 \qquad (3.10)$$

The *EMF* is proportional to the magnetic field (flux lines) and the particle velocity (rate of change of flux lines), which makes it proportional to the square of the particle velocity. A simplified model is provided by the script "EM_Induction". A charged particle crosses the centerline of a rectangular loop of conductor at a user supplied height and velocity. Non relativistic motion is assumed and low velocity forms for the fields are used. Plots of the EMF induced by the moving charge appear in Figures 3.27 and 3.28 for a velocity of two and heights of 0.2 and 0.5 above the loop, where the loop has area one, respectively. Note the reduced value of the EMF and the

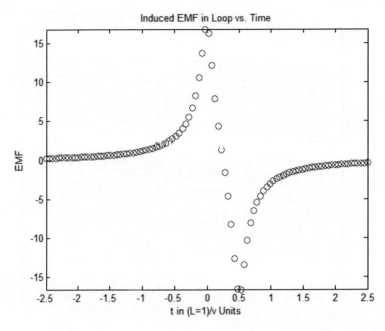

Figure 3.27: Induced EMF for a charged particle crossing the center of a conducting loop with velocity 2 and at a height 0.2 above the loop.

wider time structure for the case of a height of 0.5 compared to a height of 0.2. The EMF appears as a movie tracking the motion of the charge.

The user can establish the velocity dependence of the induced EMF through the user supplied velocity in the dialogue and by viewing the resulting plots which are movies with equal time difference between the frames shown as blue *o*.

3.14. Magnetic Bottle

For many years physicists have designed "magnetic bottles" which use shaped magnetic fields to contain a hot plasma in a search for fusion containment. In this way, the high temperatures needed to initiate and control a fusion reaction for a plasma fuel are to be achieved in a safe fashion. The latest attempts favor magnetic confinement over other techniques such as inertial confinement.

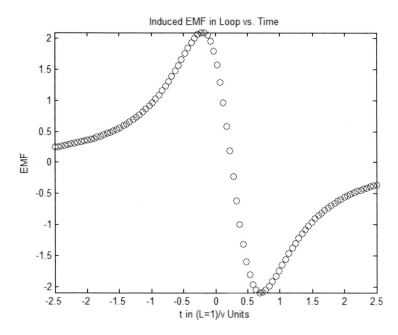

Figure 3.28: Induced EMF for a charged particle crossing the center of a conducting loop with velocity 2 and at a height 0.5 above the loop.

A test reactor with magnetic confinement (a "tokamak") is being built in France with the research goal of producing net fusion power.

The script "Mag_bottle" makes a very simplified three-dimensional model of such a bottle. The basic equation is the Lorentz force, $\vec{F} = q(\vec{v}x\vec{B})$. The fields are chosen to "reflect" the charged particles from regions of high field gradient. The radial field goes as the hyperbolic cosine of the radius, while the z field approaches $[z]$ far from the origin. The fields appear in Figure 3.29.

The equations of motion for a charged particle in an arbitrary magnetic field are solved using "ode45" and the positions and velocities are displayed using "plot3". For the specific case of initial positions, $[x \quad y \quad z] = [0.5 \quad 0 \quad 0]$ and velocities, $[v_x/v_y v_z] = [0 \quad -0.2 \quad 2]$ the last frame of the trajectory movie which is provided is shown in Figure 3.30.

Note that the particle tends to be "reflected" from the high field regions. In this case the charged particle is contained for this time

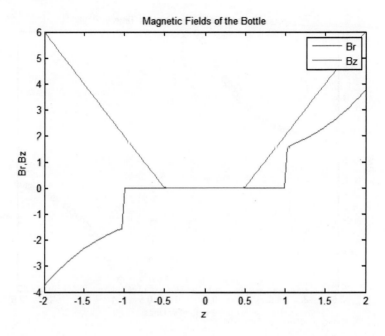

Figure 3.29: Radial and axial fields of the bottle used as an example.

interval. However, the user is encouraged to vary the parameters in order to see how "leaky" the bottle is. If it were easy the world would have a many clean fusion reactors operating for some time by now. For example, the same initial positions and with initial velocities $[0, -0.4, 3]$ the particle velocities appear in Figure 3.31 and they imply that the charge will escape.

3.15. A Fusion Reactor

The sun is a fusion reactor, where the plasma is contained by gravity. The required temperatures are very high for high fusion rates because the thermal energies must overcome the Coulomb repulsion of the charged reaction products. Therefore, confinement on earth must be accomplished by non-material means, such as magnetic fields. We simply assume that a plasma is confined in a magnetic bottle and the temperature is raised by external electric

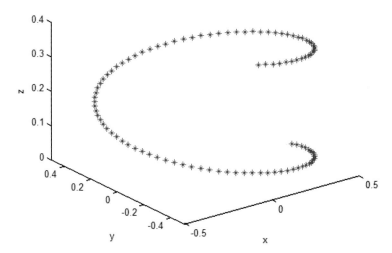

Figure 3.30: Trajectory for a specific example of a particle in a "magnetic bottle".

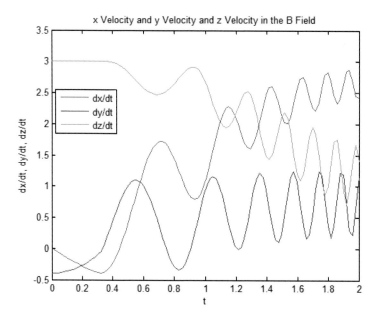

Figure 3.31: Velocities in the magnetic bottle for initial positions, $[0.5, 0, 0]$ and velocities $[0, -0.4, 3]$.

fields. The script appears in "Fustion_ITER" where ITER refers to an experimental fusion reactor being built in France.

The normalized Maxwell–Boltzmann proabability, dN/dE, is given in Eq. (3.11) where the plasma energy is E and the plasma temperature is T.

$$\frac{dN}{dE} = \frac{2\sqrt{\frac{E}{\pi}}[e^{-E/kT}]}{(kT)^{3/2}}$$
$$\frac{dr}{dE} = \frac{n_D^2 v_D \sigma_{DD}\, dN}{dE} \qquad (3.11)$$

The reaction rate density, r is proportional to the square of the deuterium plasma density, the deuterium velocity and cross section. Each plasma particle can react with n_D others, which explains the square dependence of the total rate. The cross section sets the reaction probability scale and the velocity is needed to initiate a reaction. Overall the energy dependent cross section is weighted by the distribution of plasma energies, as set by the Maxwell–Boltzmann factor.

The results of a dialogue initiated by this script appear in Figure 3.32.

```
>> Fusion_ITER
   Power density in fusion reactions with hot plasma

   Look at deuterium fusion followed by tritium
   D + D -> T + p (3.6 MeV)
   D + T -> 4He + n (17.6 MeV)
   n + 7Li -> 4He + T + n
   Enter the value of plasma temperature in 10^9 Degree Units : 0.1
   Enter the plasma density in deuterons x 10^22/m^3 : 1
   Warning: Maximum function count exceeded; singularity likely.
   > In quad at 103
     In Fusion_ITER at 73
   Total Fusion Rate, #/m^3*sec = 6.65664e+18
   Total Fusion Power Density, MWatts/m^3 = 13.2068
```

Figure 3.32: The printout made by the script "Fusion_ITER" showing the exothermic reactions. The user chooses the temperature and density of the deuterium plasma.

The reaction rate per unit volume is computed by using the MATLAB facility "quad" to integrate over the spectral reaction rate which is formed from the Maxwell–Boltzmann energy distribution convoluted with the $d+d$ (d denotes deuterium) cross section energy dpenendence. The $d+d$ reaction cross section is roughly constant at about 0.001 barns for energies above about 20 keV. The reaction rate as a function of deuterium energy is shown in Figure 3.33. Also shown, red *, is the mean energy at the user defined temperature, $\langle E \rangle = 3kT/2$. It is clear that the maximum of the reaction rate occurs at a higher temperature due to the sharp rise in the d-d cross section with energy.

The reaction rate for d-d fusion as a function of both temperature, which sets the thermal energy distribution, and energy, which reflects the reaction cross section, appears in Figure 3.34. Clearly the fusion rate increases very rapidly with energy, which puts a premium

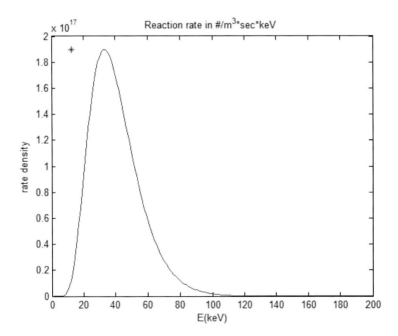

Figure 3.33: Reaction rate density as a function of d-d $C.M.$ energy. The peak is at about 50 keV, substantially above the mean energy at the chosen temperature, shown as a red ∗.

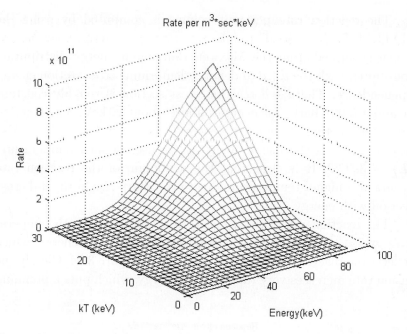

Figure 3.34: Reaction rate density as a function of d-d C.M. energy. At high temperatures, the rate increases very rapidly.

on achieving the highest contained plasma temperature possible. For example, a 100 million degree plasma with a density of 10^{22} deuterons per cubic meter would generate thirteen megawatts in a cubic meter of fuel. A factor of two more temperature raises that to about one hundred watts per cubic meter. That is a truly impressive power density.

3.16. Dipole Radiation

Dipole electromagnetic radiation is explored in the regime where the radius and inverse wave vector, r and $1/k$, are much larger than the size of the dipole in the script "Dipole_Power". The velocity c is taken to be one. The expression for the dipole power angular distribution is shown in Eq. (3.12). The dipole angular distribution is the sin squared of the polar angle of k with respect to the dipole direction. There is a wave outgoing at the speed of light which falls as a radiated energy as inverse of radius, so that the power crossing a sphere of radius r is independent of the size of r. This is the basic characteristic of a radiation field.

$$\frac{dP}{d\Omega} = \frac{k\sin^2\theta\sqrt{1+\left(\frac{1}{kr}\right)^2}[\cos(k(t-r)) + \tan^{-1}(kr)]}{r} \quad (3.12)$$

A movie of the power is provided in Figure 3.35 where the outgoing contour of the power in the (x, y) plane is shown. The dipole is assumed to be oriented along the y axis. Contours of constant power appear in a movie as a function of time for the dipole oscillations which drive the radiation. The radiative nature of the power is quite evident.

A second movie for the radiated power appears in Figure 3.36. In this case the power is plotted as the height above the (x, y) plane. There is a trough at y of zero due to the dipole nature of

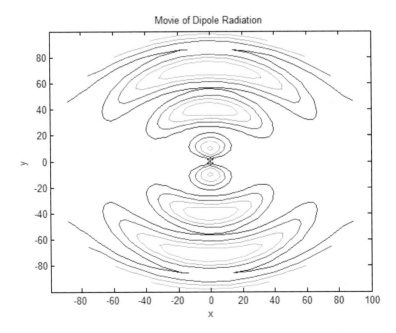

Figure 3.35: Radiated angular pattern for a dipole when r and $1/k$ are greater than the size of the dipole itself.

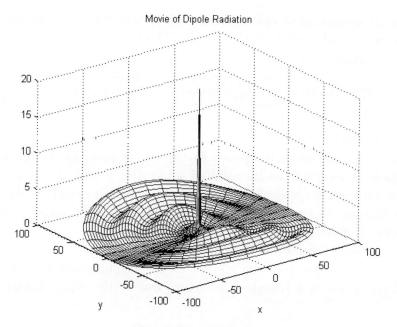

Figure 3.36: Radiated angular pattern for a dipole when r and $1/k$ are greater than the size of the dipole itself. The wave crests move out with the velocity of light and reflect the dipole oscillation frequency, seen here in red.

the radiation. The outgoing wave crests move at the speed of light and the frequency of the waves is defined by the oscillations of the dipole moment, which are visible near the origin.

3.17. Two Antennas

The relative phase of two antennas and the distance between them allow the designer to set up the enhanced directionality of the overall power with respect to the angular distribution of a single dipole antenna. This ability can be useful when it is desired to "beam" a signal to a specific direction. The electric field, expressed in Eq. (3.13), has a phase set by the wave vector, k, the distance between the two quarter wave antennas, d, located at $\pm d/2$ on the x axis and the relative phase of the driving function for the two antennas, δ. The resultant is a radiative solution, so that the electric field times radius, r, has no r dependence

$$r|\vec{E}| = \cos\left(\frac{\pi \cos\theta}{2}\right) * \frac{(e^{i\delta_1} + e^{i\delta_2})}{\sin\theta}$$

$$\delta_1 = \frac{kd\sin\theta\cos\phi}{2}, \quad \delta_2 = \frac{kd\sin\theta\cos\phi}{2+\delta}$$

(3.13)

A contour of the field is shown in Figure 3.37 for the case when the script "Two_Dipoles" is run with $kd = 5$ and zero relative phase. The shape is shown in the surface plot in Figure 3.38.

A surface plot with a choice of $kd = 5$ and 3 radians of relative phase is shown in Figure 3.39. The direction of the resultant radiation is indeed changed very significantly, thus illustrating how one can beam a signal by controlling the phase between the driving power of the two antennas.

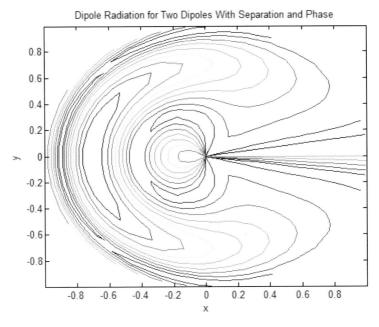

Figure 3.37: Contour plot of the resultant electric field for two antennas with spacing, $kd = 5$, and no relative phase. The pattern is distinctly not a simple dipole pattern.

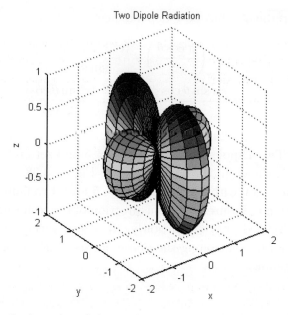

Figure 3.38: Surface plot of the resultant electric field for two antennas with spacing, $kd = 5$, and no relative phase.

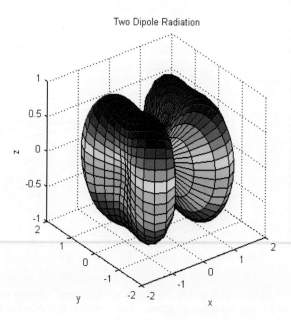

Figure 3.39: Surface plot of the resultant electric field for two antennas with spacing, $kd = 5$, and a relative phase of 3 radians.

3.18. Dispersion

Waves often encounter dissipative forces such as friction or viscosity. In those cases the wave equation acquires an added term proportional to the first partial derivative with respect to time, analogous to velocity dependent forces in the study of ballistic motion or rocket drag.

$$\frac{d^2y}{dt^2} + \delta\frac{dy}{dt} = c^2 d^2 \frac{y}{dx^2}$$
$$y = Xe^{i\omega t}, \quad X = Ae^{\lambda x} + Be^{-\lambda x}$$
(3.14)

The solutions, by separation of variable, are exponentials in space and time. The imaginary part corresponds to a traveling wave with a velocity, v, no longer c, but with a velocity which is frequency dependent. There is also a real part corresponding to attenuation, which is also frequency dependent. Note that, as δ approaches zero, the non-dispersive wave solutions, with no attenuation and velocity of c, are recovered. The attenuation is μ which is zero for δ of zero, as expected, and is approximately the imaginary part of the propagation vector, k. The wave travels in x with velocity v and circular frequency ω. Note that with larger δ values the velocity will decrease and the attenuation will increase. Larger frequencies also mean increased attenuation.

$$\mu = \frac{\omega}{c\sqrt{\frac{\left(\frac{\delta}{\omega}\right)^2}{2}}}$$

$$v = \frac{c}{\sqrt{1 + \frac{\left(\frac{\delta}{\omega}\right)^2}{2}}}$$
(3.15)

$$y \sim Ae^{-\mu x} e^{i\omega(t-x/v)}$$

The solutions are explored in the script "Dispersion2". The user chooses both the constant δ and the circular frequency ω. A movie of the traveling wave is produced and the first and last frames are superimposed in Figures 3.40 and 3.41. The velocity for the choice of parameters is only 58% of that in the case without dispersion. Light

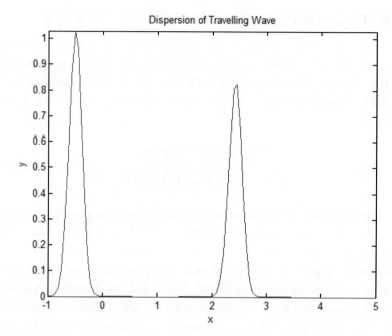

Figure 3.40: Superimposed movie frames, first red, last blue of the travelling wave in the case of $\delta = 1$ and $\omega = 0.1$, showing minor attenuation.

would have travelled to x of five at the end of the movie without dissipation.

The travelling waves are shown just to illustrate the attenuation and the velocity changes that occur with a change of the parameter δ. The wave is monochromatic. For a wave packet made up of waves of different frequencies, the differing attenuation and velocity of the different frequency components would distort the initial packet with time.

3.19. Waveguide

Waveguides are metallic structures in which radio frequency electromagnetic radiation propagates. For rectangular hollow conductors of transverse size an in x and b in y with waves traveling along z the

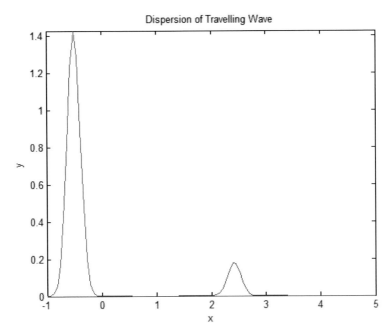

Figure 3.41: Superimposed movie frames, first red, last blue of the travelling wave in the case of $\delta = 1$ and $\omega = 1$. The velocity is the same as Figure 3.40, but the attenuation is increased.

electric and magnetic fields in z are:

$$B_z = B_o \cos\left(\frac{m\pi x}{a}\right) \cos\left(\frac{n\pi y}{b}\right)$$
$$E_z = 0 \tag{3.16}$$
$$\frac{v_{cut}}{c} = \sqrt{\left(\frac{m}{2a}\right)^2 + \left(\frac{n}{2b}\right)^2}$$

Here it is assumed that the fields are on the inner surface and inside the conductor. There is a cutoff velocity, c_{ut} below which no travelling wave can exist. The modes are defined by the integers (n, m) one of which must be greater than zero. The transverse electric mode has the lowest cutoff frequency and it is the most common mode employed in waveguides. For $m = 1$ and $n = 0$ the lowest frequency possible is $\omega > \pi c/a$.

In the lowest transverse electric mode the z and x electric fields are zero as is the y magnetic field. The spatial dependence of the non-zero fields is:

$$B_z = B_o \cos\left(\frac{\pi x}{a}\right)$$

$$B_x = \frac{-ika}{\pi}\left[B_o \sin\left(\frac{\pi x}{a}\right)\right] \qquad (3.17)$$

$$E_y = \frac{-i\omega a}{\pi c}\left[B_o \sin\left(\frac{\pi x}{a}\right)\right]$$

The transverse electric modes in a rectangular waveguide appear in the script "Waveguide_TE". The non-zero fields in this mode appear in Figure 3.42 for the users choice of aspect ratio, a/b, of one.

The lowest possible frequencies in the different modes appear in Figure 3.43. Basically, the waves must satisfy the boundary

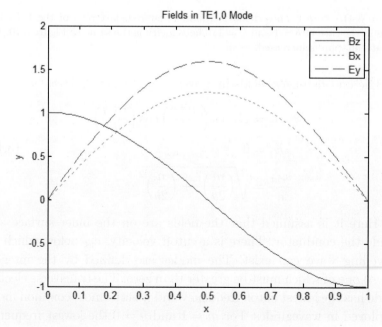

Figure 3.42: Plots of the transverse, (x, y) dependence of the non-zero fields in the case of the lowest (1,0) TE mode for $a/b = 1$.

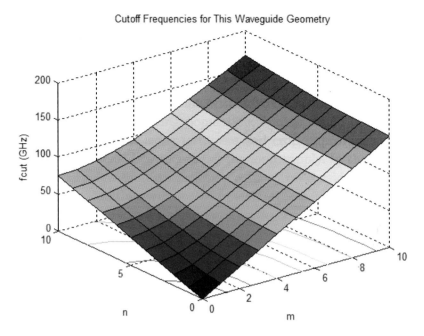

Figure 3.43: Cutoff frequencies and contours in the TE mode for a specific, user chose, a/b aspect ratio of the waveguide, in this case a/b of one. Modes up to (10,10) are plotted.

conditions and a minimum frequency is needed if the wave is to oscillate from one wall to the wall on the opposite side. Larger mode numbers mean that more oscillations are needed which, in turn, means higher cutoff frequencies. In the figure a specific choice of aspect ratio, a/b of one was made and the case was set by a waveguide of one cm, which has a TE cutoff of 15 GHz.

3.20. Skin Depth

Previously it had been assumed that the fields inside a "perfect" conductor do not penetrate the conductor. That is true in the case of static fields for perfect conductivity. However, an electromagnetic wave exists in a conductor although it is damped strongly.

The conductivity σ relates the current density, J, and the electric field, E, in the microscopic form of Ohm's law. The wave in the conductor has a complex wave vector, k, which means that there

```
>> Skin_Depth
   Skin Depth - EM Wave in a Conductor


   EM Wave Tunnels into a Conductor
   Enter Wave Frequency in MHz : 100
   Skin Depth in um = 6.5
```

Figure 3.44: Printout for a specific user input specifying the two important parameters, material and frequency.

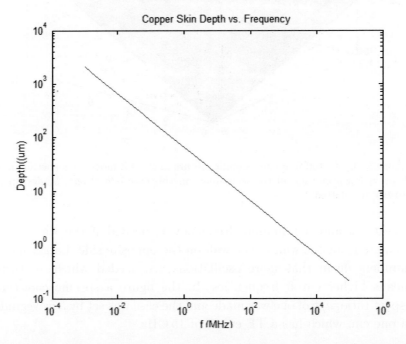

Figure 3.45: Skin depth for copper as a function of the frequency of the incident wave.

is an exponential penetration of the wave into the conductor by a characteristic distance d which is proportional to the inverse of the imaginary component of the wave vector k. The form for k in Eq. (3.18) is closely related to the previous discussion of dispersion, with σ playing the role of the parameter δ in Eq. (3.14)

$$\vec{J} = \sigma \vec{E}$$
$$k^2 = \left(\frac{\omega}{c}\right)^2 \left[1 + i\left(\frac{4\pi\sigma}{\omega}\right)\right] \qquad (3.18)$$
$$d \sim \frac{c}{\sqrt{2\pi\omega\sigma}}$$

This effect is called the "skin depth" for the conductor. The script which looks at this effect is called "Skin_Depth". The user is invited to choose the conductor and the frequency of the incident wave. The printout of the script appears in Figure 3.44.

A plot for copper over a wide range of frequency is also provided as shown in Figure 3.45. It is clear that high frequency waves populate only the "skin" of the conductor with typical penetration distances in microns. This is why waveguides for MHz waves and higher frequencies need not be made of thick metal. On the other hand, 60 Hz home wiring has fields that penetrate centimeter distances with resulting implications for the hardware.

Chapter 4

Gases and Fluids

"Big whorls have little whorls, which feed on their velocity, and little whorls have lesser whorls, and so on to viscosity."
— **Lewis Richardson**

"Turbulence is the most important unsolved problem of classical physics."
— **Richard Feynman**

4.1. Leaky Box

An interesting problem in statistical mechanics is the behavior of a container of gas molecules with an aperture through which a gas molecule may escape. As is intuitively obvious, faster molecules have a larger chance to escape, so that the gas remaining in the container cools as time goes on. The energy flow out of the box depends on temperature T, average velocity, $\langle v \rangle$ and the aperture size A, so that the flow should be proportional to $kT\langle v \rangle A$.

In three dimensions the number of molecules per magnitude of velocity dn/dv goes as the square of v, so that the number of wall collisions goes as the cube of v. The model in two dimensions is supplied in the script "Box_Leak". A two-dimensional Maxwell–Boltzmann energy distribution is a simple exponential in the energy. Particles are picked out of this distribution, using the Monte Carlo technique explored earlier, and then their momentum is given a random direction. The initial positions are randomly assigned inside the box.

A specific example appears in Figure 4.1. The user chooses the temperature, the size of the box and the aperture. In that way one can confirm that in this simple model, the total number of collisions with the walls depends on the temperature and that the fraction escaping through the aperture does also. The printout shows that the momentum of those molecules exiting the aperture, is, on average,

```
Mean Momentum Remaining 0.391165
Input the Number of Gas Molecules : 100
Input the Number of Time Steps : 100
Input the Gas Temperature * k, Mass = 1: 0.2
Mean Velocity 0.535365
Input the Box Height, Length = 1: 2
Input Box Hole Size: 0.4

Number of Wall Collisions 246

Momentum Impulse to Walls 244.53

Number of Leaked Molecules 25

Total Lost Momentum 15.6369

Total Initial Momentum 53.5365

Mean Momentum Leaked 0.625475

Mean Momentum Remaining 0.505329
```

Figure 4.1: Printout of a user choice for "Box_Leak". The reduction in the mean momentum of the molecules remaining in the box is clear.

higher than those molecules that remain in the box. A frame of the movie in a specific example appears in Figure 4.2. In this case about one-third of the molecules escape during the one hundred time steps over which the molecules are tracked. Other parameter choices should be tried by the interested user.

4.2. Rectangular Flow

In looking at flow in a rectangular pipe the boundary conditions are that the fluid has zero longitudinal velocity at the transverse boundaries. The flow is driven by a pressure drop, ΔP of a fluid with viscosity η over a length L. The Fourier coefficients for different modes defined by n and m require that m and n are odd to satisfy the boundary conditions at the rectangular walls. The Fourier coefficients are defined by $C_{m,n}$ in Eq. (4.1) for a given choice of modes and a given geometry.

The velocity is driven by the pressure drop, but falls with distance along the pipe, set by L, due to the dissipative properties of the

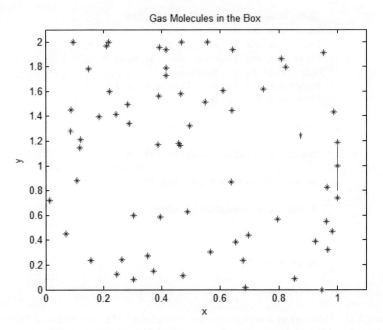

Figure 4.2: Plot of the "gas" molecules at the end of 100 time steps with 100 initial molecules. The aperture is in red and the escaped molecules are pinned to the center of the aperture. The mean momentum for particles remaining in the box is reduced compared to the initial mean, while the mean of the escaped particles is increased.

viscosity. The pipe is square, size a on a side

$$\frac{\partial^2 v_z}{\partial x^2} + \frac{\partial^2 v_z}{\partial y^2} \sim \frac{\Delta P}{\eta L}$$

$$v_z(x,y) = \sum_m \sum_n \sin\left(\frac{n\pi x}{a}\right) \sin\left(\frac{n\pi y}{a}\right) C_{m,n} \qquad (4.1)$$

$$C_{m,n} = \frac{nm}{\left[\left(\frac{n}{a}\right)^2 + \left(\frac{m}{b}\right)^2\right]}$$

The behavior of specific modes is provided by the script "Flow_Rect2". The velocity contour for the lowest mode appears in Figure 4.3 while the velocity profile for $n = 1, m = 3$ appears in Figure 4.4. The user can specify n and m and explore the higher order modes in this fashion.

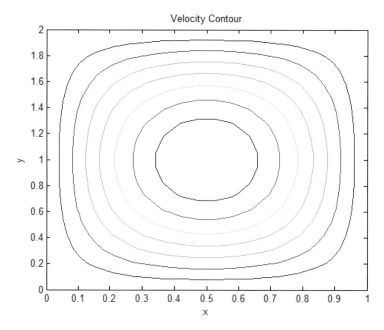

Figure 4.3: Velocity contour for the $n = 1$ and $m = 1$ mode. The velocity is monotonically decreasing from the center of the pipe and vanishes on the boundaries.

4.3. Debye Temperature

Classically, the specific heat at constant volume, C_V, of a solid is a constant, in contradiction to the experimental data at low temperatures. The classical prediction is about 25 Joules per degree Kelvin. Einstein first achieved a partial solution by quantizing the "oscillators" that he assumed made up the solid. However, the low temperature behavior of the resulting prediction did not reproduce the data. At low temperatures the specific heat of solid experimentally goes as the cube of the temperature.

Debye used Bose-Einstein statistics to treat the "phonons" or quantized thermal oscillations of the atoms in the solid. In contrast to Einstein, Debye allowed modes with several atoms participating rather that the Einstein approach of independent atomic oscillators. The resulting theory has a spectral energy density, $du(E)/dE$ with

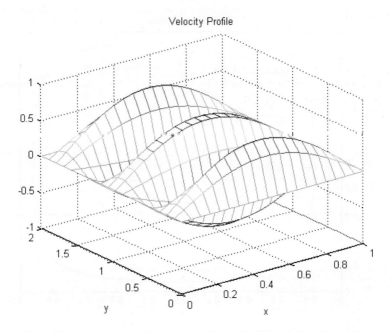

Figure 4.4: Velocity contour for $n = 1$ and $m = 3$. The plot peaks at the center of the pipe and vanishes on the boundaries but has intermediate structure as opposed to the lowest (1, 1) mode.

a similar dependence as the black-body Stefan-Boltzmann theory, $du(E)/dE \sim E^3/(e^{E/kT} - 1)$ but has a parameter which must be measured, the Debye temperature, T_D. Having chosen it both the low and high temperature behaviors are in agreement with experimental facts. The expression for the specific heat in this case is found by differentiating $u(E)$ and is:

$$C_V = 9kN \left(\frac{T}{T_D}\right)^3 \int_0^{T_D/T} \frac{x^4 e^x \, dx}{(e^x - 1)^2} \qquad (4.2)$$

Printout from the script "Debye" is shown in Figure 4.5. Both the Einstein and Debye expressions for specific heat are computed and plotted. The integral expression in the Debye case is evaluated numerically using the MATLAB utility "integral" since the expression does not have a closed form symbolic solution.

```
>> Debye
   Specific Heat of Solids - Einstein and Debye


Classically Cv = specific heat = 3kN = 24.9 J/K
  Experimentally, Cv goes as T^3 at Low Temp  kTe = Einstein Energy
  Input the Einstein Energy - kT = 1/40 eV @ 300 K : 0.03
  Input the Debye Energy - kT = 1/40 eV @ 300 K : 0.03
  Debye Temperature for Cu is 344 oK
```

Figure 4.5: Printout from the script "Debye" with the user defined temperatures for the Einstein and Debye specific heats.

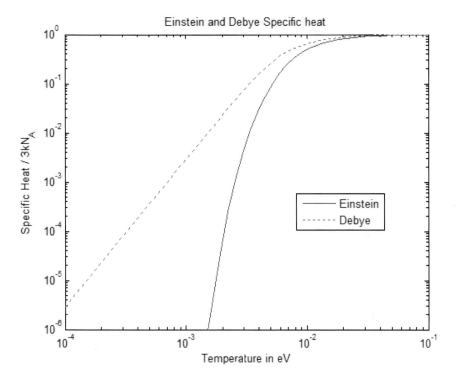

Figure 4.6: Einstein and Debye specific heat as a function of the absolute temperature of the sample expressed in thermal energy units.

The specific heat vs. temperature for the two models is shown in Figure 4.6. Note the rapid falloff of the specific heat in the Einstein case, while the Debye theory displays the power law behavior of the experimental data in this log-log plot.

4.4. Times Arrow

Entropy for a closed system always increases or stays the same. The earth is not closed and we can violate this entropy increase because the sun supplies us with energy. In the case considered here a closed box of gas molecules is used to illustrate the principle of entropy increase. A "gas" is prepared to be localized in the bottom half of a container in the script "Times_Arrow2". The user chooses the number of molecules, the number of time steps to track them, the temperature of the molecules and the height of the container. Note that the energies of the molecules are given a two-dimensional Maxwell-Boltzmann distribution in energy using Monte Carlo methods. The initial box for a specific example, 100 molecules, 100 time steps, and temperature of 0.01 and box height of one is shown in Figure 4.7.

The distribution of molecules after one hundred time steps is displayed in Figure 4.8. The number in the top half is now 36, while 64 remain in the bottom. The movie which is provided allows the

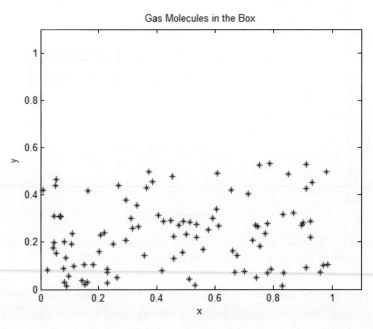

Figure 4.7: Initial distribution of molecules confined to the lower half of a container.

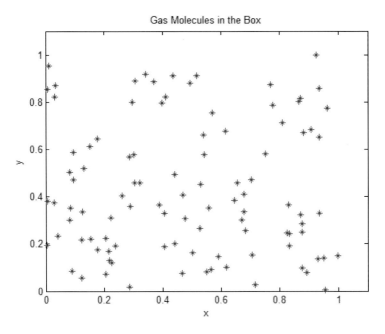

Figure 4.8: Distribution of 100 molecules in a container after 100 time steps.

user to follow the diffusion of the molecules. The rate of diffusion is controlled by specifying the temperature. It is quite clear that it is highly unlikely that the molecules will later all once again appear in the bottom of the container. The entropy of the final state of the system is clearly larger than that of the initial state, even though the basic physics is time reversible. Indeed the laws of microscopic physics obey time reversal. For example, a Keplerian orbit would be the same if time ran "backwards". It is in complex, many particle, systems that entropy becomes evident.

4.5. Compressibility

The compressibility of materials can be related to the fundamental structure of the solid. A simple case is that of an ionic crystal where the positive ions have a regular structure, with a one-dimensional separation R_o. Overall ionic neutrality is assumed and a simplified one-dimensional model is used. The ionic cores repel strongly with an inverse power of separation with exponent n. The electron-ion attraction falls off much more slowly, with an energy, E, which goes as the inverse of the separation. The electromagnetic interaction involves several ionic cores due to the long range of the interaction.

$$E \sim \frac{a}{R^n} - \frac{be^2}{R}$$

$$b = 2\left(\frac{1}{R_o} - \frac{1}{2R_o} + \frac{1}{3R_o} - \frac{1}{4R_o} \cdots\right)$$

(4.3)

The Coulomb part has a sumable series, $b = 2*\ln(2)$. The minimum of the energy, E_o, for a given ion occurs when E is minimized with respect to R:

$$E_o = \frac{be^2 \left(\frac{1-1}{n}\right)}{R_o}$$

(4.4)

The relationship of the crystal structure and the compressibility of the material is explored in the script "Compress". Printout from the script is shown in Figure 4.9. Small radii result in a larger compressibility. A value of exponent $n = 10$ is assumed in the script for the short range radial dependence of the core repulsion.

Experimentally the compressibility, K, is a macroscopic quantity defined to be the fractional change in volume per applied pressure at constant temperature. It is also known as the bulk modulus. Typical values are approximately 10^{11}/Pa. At low temperature the change in internal energy dU is $-P\,dV$, so that dP/dV is the second derivative

```
>> Compress
   Program to look at ionic crystal - and compressibility

   Ionic crystal - core repulsion + electromagnetic attraction
   Enter the crystal radius in A: 3
   Compressibility modulus ~ 1/Ro^4 ~ 10^11 Pa

   Compresibility modulus in Pa = 1.9988e+10
```

Figure 4.9: Printout from the script "Compress." The user chooses an equilibrium radius in Å.

of U with respect to V and V goes as R^3. The value of $1/K$ for the material is called the modulus and has dimensions of pressure.

$$\begin{aligned}\frac{1}{K} &= \frac{1}{18NR_o}\left(\frac{d^2U}{d^2R}\right)_o \\ &= \frac{(n-1)be^2}{18R_o^4}\end{aligned} \quad (4.5)$$

The value of R_o can be obtained by x-ray scattering patterns. The compressibility can then be predicted for different n values. It is amusing that a macroscopic measure of K allows for the measure on the microscopic core repulsion power law n. The energy as a function of R is shown in Figure 4.10. There is a clear minimum where the core repulsion is balanced by the Coulomb attraction of the nearest neighbors, diluted by the next to nearest neighbors.

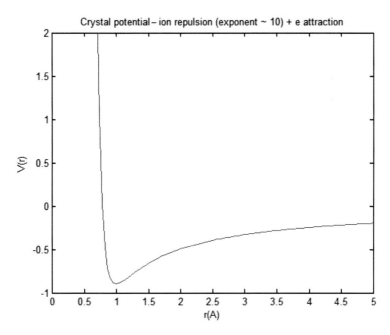

Figure 4.10: Potential energy for an ionic crystal showing the core repulsion at small R and the Coulomb attraction at large values of radius.

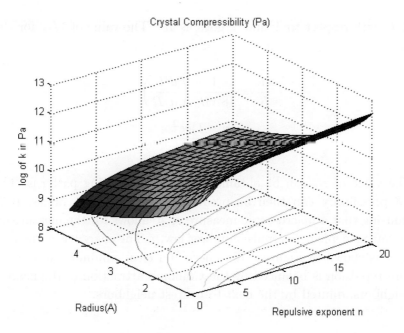

Figure 4.11: Contour of the modulus in Pa as a function of n and R. The modulus increases with n and decreases with R.

The modulus, $1/K$, is shown in Figure 4.11 as a function of the separation of the crystal ions and the repulsive exponent. Experimentally the radius is typically about 3 Å, while the exponent is about 10. The modulus, plotted logarithmically, increases rapidly with decreasing radius and increases rapidly with the value of the repulsive index. The magnitude of the modulus, $1/K$, is about 10^{11}/Pa as expected.

4.6. Viscosity

Viscosity, η, is a measure of resistance to motion through a medium. In ballistics, it leads to a terminal velocity when acceleration becomes zero. It can be measured, for example, by measuring the terminal velocity and also be several other means. It depends on temperature through the mean velocity of the medium, $\langle v \rangle$, and it depends of the diameter, d, and mass, m, of the objects which make up the medium. That means knowing the viscosity, the size, diameter d, of the molecules in a gas can be indirectly measured

$$d^2 = m \frac{\langle v \rangle}{3\pi\eta} \qquad (4.6)$$

A typical value of viscosity in air is $18.1/\mu$Pa*sec, where one atmosphere at STP is 0.1 Pa. The mean thermal velocity is assumed to be $3kT/2$. Taking the atmosphere to be all nitrogen N_2, Å is 28. The mean velocity is then 0.51 km/sec. In this case, the molecules are inferred to have a diameter of 3.7 Å. Again, a measured macroscopic quantity, the viscosity, allows for a fundamental determination of the diameter of a molecule.

Printout from the script "Viscosity_Molecule" is given in Figure 4.12. The user supplies the atomic weight of the air given the measured air viscosity. The temperature is assumed to be STP. In Figure 4.13 the apparent size dependence of the molecules is shown as a function of kT. The dependence is weak, since d scales as the square root of T.

```
Program to look at relationship of viscosity and molecular diameter

Gas viscosity depend on mass, mean velocity and size of molecule
Measure gas viscosity, eta , -> find size of molecules

m =

2*A*mp


v =

((3*kT)/(2*A*mp))^(1/2)

|
d =

((2*A*mp*((3*kT)/(2*A*mp))^(1/2))/(3*eta))^(1/2)/pi^(1/2)

Enter atomic weight of gas atom: 1
Mass of molecule, mean thermal velocity and diameter
molecular mass = 3.34e-27 (kg), thermal velocity = 1895.47(m/sec) and size = 1.93179 (A)
Enter atomic weight of gas atom: 14
Mass of molecule, mean thermal velocity and diameter
molecular mass = 4.676e-26 (kg), thermal velocity = 506.586(m/sec) and size = 3.73674 (A)
```

Figure 4.12: Printout from the script "Viscosity_Molecule" for two choices of Å made by the user.

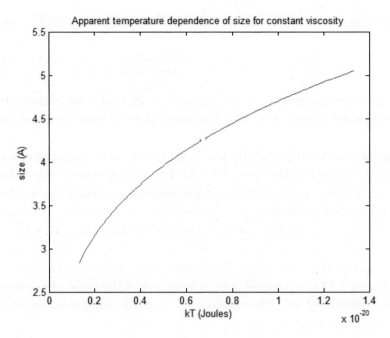

Figure 4.13: Change in apparent molecular size with temperature. A slow square root dependence on T is shown.

4.7. Water Waves

Water waves are explored for waves in deep water, where the relationship of the frequency of the wave to the wave vector is $\omega(k) = \sqrt{gk}$ or in shallow water where $\omega(k) = k\sqrt{gh}$ and h is the water depth. The expression for the incremental wave height z due to an initial disturbance, $A(k)$, as a function of r and t is:

$$z(r,t) = 2\text{Re} \int_0^\infty A(k) k r(k) J_o(kr) e^{-i\omega t} \frac{d\vec{k}}{2\pi} \qquad (4.7)$$

The water waves are studied in the script "Water_Waves" which employs the MATLAB functions "besselj" and "int" to perform the necessary integration. The amplitude, $A(k)$, of the initial wave is a sharply peaked Gaussian in k^2, which gives an approximate delta function in r at t equal to zero as an initial condition. The dispersion of water waves is taken into account by the relationship shown in

Eq. (4.8), where the frequency approaches the deep water result without dispersion for long wavelengths, kh much less than one.

$$\omega(k) = \sqrt{gk\tanh(kh)} \qquad (4.8)$$

In the script a sharply peaked Gaussian initial wave at small radius is generated. The wave is then followed out in radius r and time t and a movie of the motion is created. Results for shallow water height, no dispersion, as a function of radius for the last movie frame are shown in Figure 4.14.

The script allows for the user defined choice for shallow water, with and without dispersion and for deep water. The deep water result, the last movie frame of wave height vs. radius, is shown in Figure 4.15. The wave velocity in this case differs from the shallow water case and the wave is less peaked. The mesh plot of the deep water wave, again last movie frame, is shown in Figure 4.16. Dispersive results can also be generated by the user as desired.

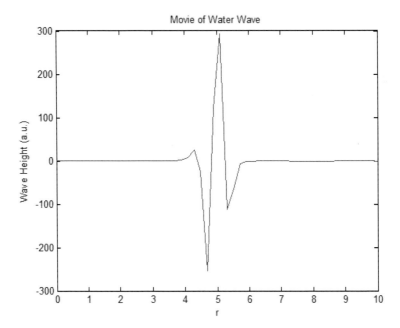

Figure 4.14: Last frame of a movie for a shallow water wave, $h = 0.1$ m, without dispersion.

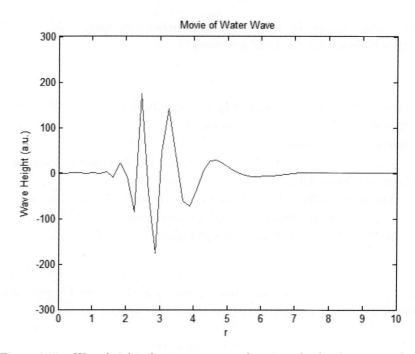

Figure 4.15: Wave height of water wave as a function of r for deep water; last frame of a movie. The velocity is less than that of Figure 4.14 and the radial structure is more oscillatory.

4.8. Semiconductor

The use of semiconductors in the electronics industry is ubiquitous. A very small introduction to the topic appears in the script "Semiconduct". By intrinsic, is meant a perfect crystal of silicon or germanium without crystal defects and without any impurities. In this case there are, at a temperature of zero, all valence electrons in a "valence band" of allowed energies and no electrons in a "conduction band" of allowed energies the lowest energy of which is at an energy Eg higher than the top valence band energy.

At elevated temperatures, some electrons receive sufficient thermal energy to exist in the conduction band and to contribute to electrical conductivity. At zero temperature such electrons are "frozen in" the valence band by the Fermi exclusion principle since all available states in that band are filled. At low temperatures the

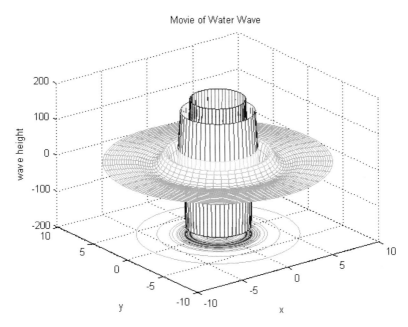

Figure 4.16: Mesh plot for a water wave in deep water. The frame corresponds to that of the previous figure, Figure 4.15. There is no dispersion and the wave crests are quite sharp.

Fermi-Dirac statistical factor is approximately, $e^{(E_F-E)/kT}$, while the density of states factor for electrons arises from the statement that all momentum components are equally probable, so that $(dP_x\, dP_y\, dP_z)/\hbar^3 = 4\pi P^2\, dP/\hbar^3$ is the density in momentum space, while in energy space the density is proportional to $(2m_e)^{3/2}\sqrt{E-E_g}\, dE/(2\pi^2\hbar^3)$.

In the approximation that the Fermi level, E_F, is at $Eg/2$ at low temperature, the number of electrons per unit volume in the conduction band is obtained by integrating the density of states over energy weighted by the Fermi-Dirac statistical factor from Eg to infinity:

$$n_e = 2\left(\frac{m_e kT}{2\pi\hbar^2}\right)^{3/2} e^{(E_F-Eg)kT} \qquad (4.9)$$

The number of holes remaining in the valence band, assuming the same effective mass for holes in the crystal as for the electrons in the crystal is:

$$n_h = 2\left(\frac{m_e kT}{2\pi\hbar^2}\right)^{3/2} e^{(-E_F)kT} \qquad (4.10)$$

Using the values for the energy gap, E_g, in silicon, 1.1 eV, and germanium, 0.7 eV, at STP a conduction band density of charge carriers is expected to be about $1.4 \times 10^{16}/m^3$ and $2.5 \times 10^{19}/m^3$ respectively. This density should be contrasted with Avogadro's number of about 10^{23}. The product of the hole and electron densities depends only on the temperature and the gap energy as can be seen from Eqs. (4.9) and (4.10). This fact will be used later in discussing a junction between two "doped" semiconductors.

It is clear that the addition of impurity states at the level of greater than $10^{18}/m^3$, if these states are near the top of the valence band or the bottom of the conduction band, but still in the gap would completely dominate the intrinsic conductivity of the crystal. Indeed, this is how the n type and p type semiconductors are created.

The script "Semiconduct" explores these concepts. Printout from the script appears in Figure 4.17. The user chooses the gap energy,

```
>> Semiconduct
   Program to look at energy levels and populations of e and holes


Intrinsic semiconductor
Enter the energy gap between CB and VB (eV): 1.1
If there are no impurity states # e = # holes, EF = Eg/2 = 0.55
# e = # holes /m^3 at 300 oK = 6.39342e+15
If intrinsic #e = #h, input new #e : 10^17
  Modified Fermi level = 0.618748 (eV)
```

Figure 4.17: Printout from the script "Semiconduct" where the user has chosen a gap energy appropriate to silicon. STP conditions are assumed.

in this case that of silicon. The addition of more electrons in the conduction band, due to added impurity states changes the Fermi level, in this case increasing it, as can be seen from Eq. (4.9). The equation shows that the Fermi energy depends logarithmically on the density of charges. The change is small in most cases, however, and a constant Fermi energy appropriate to T of zero can often be assumed.

The log of the number of intrinsic carriers is shown in Figure 4.18 as a function of temperature. The strong dependence on the gap energy indicates why germanium, with a lower gap energy than silicon, is not favored by manufacturers because the intrinsic conductivity at high temperatures competes with the doped impurity conductivity.

The dependence of the log of the intrinsic number density of electrons is shown in Figure 4.19. The strong dependence on gap

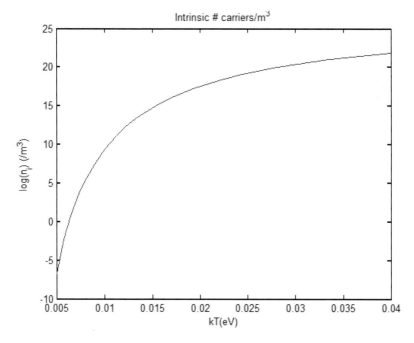

Figure 4.18: Log of the electron number density as a function of temperature for silicon.

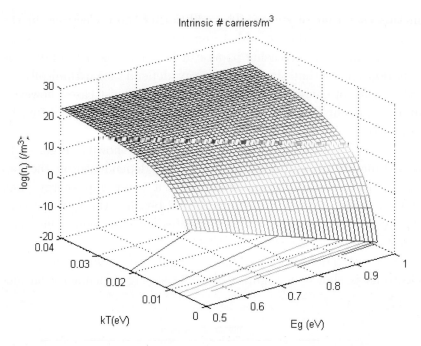

Figure 4.19: Log of the intrinsic number density of electrons for a semiconductor as a function of temperature and gap energy. STP has a kT of 0.025.

size is evident, as is the strong temperature dependence. It is evident that it can be expected that electronic devices will typically work better at reduced temperatures where the doping effects will dominate.

4.9. Semiconductor Junction

The intrinsic semiconductors can be "doped" with impurity states with energies near the bottom of the conduction band or the top of the valence band but in the forbidden gap. These states then will define the mobile charge carriers to be either electrons, n type, or holes, p type. A junction diode is a device with both types in contact. There are ions which are not mobile and there are mobile charge carriers. The Poisson equation for a region with only positive ions and both positive and negative charged carriers is shown in

Eq. (4.11) which is appropriate to the n region of the junction

$$\nabla^2 V = \frac{\rho}{\varepsilon} \sim \frac{e}{(k\varepsilon_o)P_o}$$

$$= \frac{e}{(k\varepsilon_o)}\left[P_o - 2n_o \sinh\left(\frac{eV}{kT}\right)\right]$$

$$V(0) = 0$$

$$np = n_o^2$$

(4.11)

The electron charge is e, the charge density is ρ, ε is $k\varepsilon_o$, where k is twelve for silicon, and ε_o is the vacuum permittivity. The voltage V exists can, in first approximation be defined purely by the positive ions with density P_o. Since the Fermi energies on the n and p sides of the device differ, Eqs. (4.9) and (4.10), a diffusion of charge carriers across the n/p boundary defines a dipole layer of charge at the border which acts to equalize the Fermi energies. Both the positive p and negative n charge carriers diffuse at the location x of zero, where the n/p junction occurs and where they have equal densities n_o.

Starting with the approximation of ignoring the mobile charge carriers, the field and potential in the case of full depletion, solving the Poisson equation, when the field extends from the junction throughout the diode, of full width d, is shown in Eq. (4.12). The coordinate x is zero at the junction. The field extends throughout the size of the junction at full depletion. Equal doping densities on both sides of the junction are assumed as are equal lengths of the n and p sides.

$$E(0) = E_0 = \frac{-eP_o d}{(2k\varepsilon_o)}$$

$$E(x) = -eP_o \frac{\left(\frac{d}{2}-x\right)}{(k\varepsilon_o)}, \quad x > 0$$

$$E\left(\pm\frac{d}{2}\right) = 0$$

$$V(x) = eP_o \frac{(x^2 - dx)}{(2k\varepsilon_o)}$$

$$V_d = -2V\left(\frac{d}{2}\right) = \frac{eP_o\left(\frac{d^2}{4}\right)}{(k\varepsilon_o)} = -E_o^*\left(\frac{d}{2}\right)$$

(4.12)

The field is largest at the junction at x of zero and decreases linearly to zero at the device boundary at $+d/2$ and $-d/2$ at full depletion. The potential is quadratic in x and vanishes at the origin. The printout for the script "n_p_Junction2" is shown in Figure 4.20 for a specific user set of choices. The field at x of zero is printed out as is the voltage. The device voltage is twice that.

The approximate voltage for the junction at depletion, where the mobile carriers are swept out of the device, is shown in Figure 4.21. The voltage is quadratic and vanishes at the junction. The fixed ionic charge density is also shown schematically. The specific parameters chosen are a one hundred micron device with an ion density of 5×10^{17} ions per cubic meter. The density of donor and acceptor ions is assumed to be the same.

Taking the mobile charge carriers into account requies a numerical solution to the full equation, Eq. (4.11). This is accomplished by using the MATLAB ode45 utility. The user supplies the density of mobile carriers at the junction boundary as well as the amount by

```
>> n_p_Junction2
   Program to look at a constant doped density n-p junction


Look at junction between n and p type silicon
Assume Room Temperature ~ 300 oK
Enter the full junction thickness in microns (~ 100): 100
Enter the P type ion density at junction center in 10^17/m^3 (~ 1): 5
Field at x = 0 for full depletion = -37664.8
Voltage for full depletion at x = d/2 = 1.88324
Enter the electron density at junction center in 10^16/m^3 (~ 0.01):  0.01
Enter the factor by which E(o) is reduced from full depletion:  0.5
Depth at E = 0 = 20.6646 (um), and V at that depth = 0.419733
```

Figure 4.20: Printout for a specific user chosen set of junction parameters.

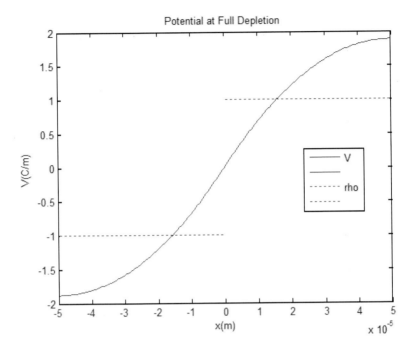

Figure 4.21: The approximate voltage at depletion for the 100 micron junction. In this case the device depletion voltage is about $2 \times 1.9 = 3.8$ volts. The left side is N type and the right side is P type — referring to the fixed ions or p type and n type, referring to the mobile carriers.

which the maximum electric field is reduced from the approximate full depletion value. That value defines the equation and sets the initial conditions for ode45, along with $V(0) = 0$. The actual density at the boundary is only estimated by the user.

Results for the electric field appear in Figure 4.22. Note that the field at small x is linear as before, and an approximate proportionality of the field at x of zero and the depletion region holds. In the example the depletion of mobile carriers by the field is not complete but extends only over about half of the full device. At large distances from the junction the density of mobile carriers apporaches the donor density. Other sets of parameters can easily be explored. The field and potential and n and p densities are calculated only for the positive x or P type region. For negative x, symmetry can be invoked.

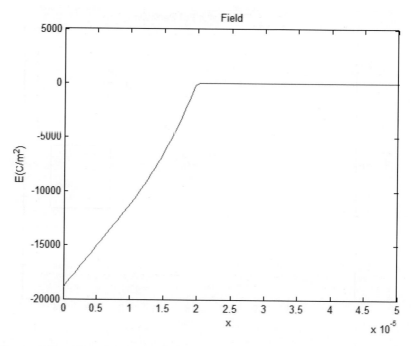

Figure 4.22: The electric field found numerically in the case where mobile carriers are taken into account. The field near $x = 0$ is linear. The parameters for the example are given in Figure 4.20.

4.10. n-p Diode

The np junction has rectifying behavior when an external voltage is applied. The diode is covered in the script "n_p_Diode" which looks at the response to an applied voltage. The dialogue for that script in a specific example appears in Figure 4.24.

The doping by donors and acceptors, P and N ions and n and p charge carriers shifts the Fermi level, as mentioned previously. The Fermi levels for a doping concentration given in Figure 4.24 are shown in Figure 4.25. Donors put electrons in the conduction band and raise the Fermi energy. The opposite is true for acceptors. The values follow from Eqs. (4.9) and (4.10) $E_{Fn} = E_g + kT \log(P_o/n_e), E_{Fn} = -kT \log(P_o/n_e)$. The diffusion of carriers at the junction creates a dipole layer which then equalizes the Fermi levels, resulting in

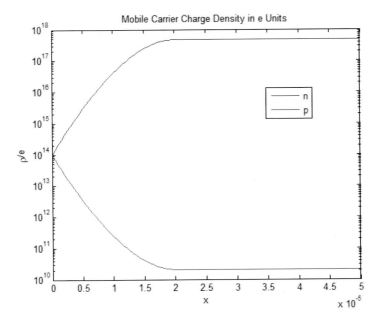

Figure 4.23: The n and p charge densities found numerically in the case where mobile carriers in thermal equilibrium are taken into account. The densities are equal at $x = 0$, but on the P side, $x > 0$, the n are about 10^7 times larger than the p density. The p are minority carriers, and the n are majority carriers in this region.

```
>> n_p_Diode
   Program to look at a constant doped density n-p junction diode


Look at junction between n and p type silicon, diode rectification
Assume Room Temperature ~ 300 oK
Enter the full junction thickness in microns (~ 200): 200
Enter the P type ion density at n side in 10^18/m^3 (~ 1): 1
Intrinsic # e = # holes = 6.39342e+15, EF=Eg/2 = 0.55
Assume fully ionized doners and acceptors, equal doping n = Po, p = Po
Modified Fermi level on n side and p side = 0.423688, 0.676312 (eV)
Dipole layer equalizes EFp and EFn, Vbi = 0.252624 (eV)
Enter the external applied voltage on n side(V): 2
Depletion Width (um) = 77.3351
Field at interface (V/m) = -58256.2
Rectifying Current for this applied voltage (a.u.) = 5.54062e+34
```

Figure 4.24: Output resulting from a dialogue with the script "n_p_Diode". The user chooses a junction thickness and doping density. An applied voltage is also chosen.

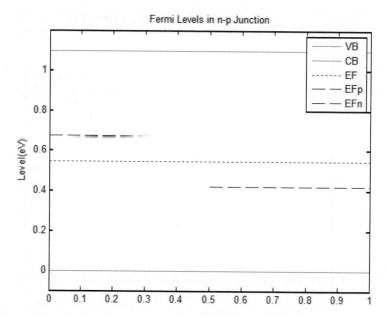

Figure 4.25: Fermi energies for the intrinsic case (dotted line) and for n and p type doping (dashed line) for silicon and doping levels as given in Figure 4.24. The bottom of the valence band (VB) and the top of the conduction and (CB) are indicated by solid lines.

a "built in" voltage V_{bi} which exists without an external applied voltage. The value for V_{bi} is just the difference in Fermi levels in the n and p sides of the diode.

The depletion region depends quadratically on the applied voltage. Away from that region, it is assumed that all the donors are singly ionized so that the n density is Po. On the other side the n density is reduced and is found by assuming equal doping, $n = n_i^2/Po$ where n_i or n_e is the intrinsic carrier density. A simple interpolation in the depleted region results in Figure 4.26. In this script, the contribution of the mobile carriers to the field and potential has been ignored, while it was approximately taken account previously in the script "n-p_Junction2".

The diode acts as a rectifier. Almost independent of voltage there exists a generated current, Ig, where holes thermally generated in the n region diffuse into the p region. Although the dipole layer

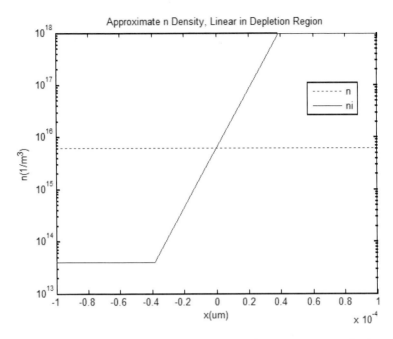

Figure 4.26: Approximate density of n carriers for a 100 micron device with 2 V externally applied. The depletion depth for this example (Figure 4.24) is 77 microns.

impedes diffusion there is also a recombination current, Ir, with the opposite sign, where holes from the p region flow to the n region and recombine. With no bias the two currents are equal and there is no net current. For forward bias, with external voltage V_o, the Fermi level is reduced for diffusing majority carriers to get through the dipole layer at the junction boundary. They are then are swept up by device and $I_r \sim I_g e^{(V_o/kT)}$. For reverse voltage applied, only the small reverse current, $-Ig$, due to the other minority carriers exists since Ir is now reduced by the exponential factor, not enhanced. The resulting idealized current versus voltage relationship is:

$$I(V) = I_g(e^{eV_o/kT} - 1) \qquad (4.13)$$

The script asks for an applied voltage. It can treat forward values, $V > 0$, and small values of reverse voltage if it is less than V_{bi}.

4.11. Freezing Pipes

The practical problem of how deep in the ground to lay in water pipes is explored in the script "Pipe_Freeze". There is an analytic solution to this problem. The heat equation is shown in Eq. (4.13) below where T is temperature, t is time and x is position. The basic parameter is a which has dimensions length squared over time which is called the diffusion coefficient. In this case a is about $2.4\,\text{m}^2/\text{year}$ or 7.6×10^{-8} for a typical soil. A basic time scale to transfer temperature by one meter is then 0.24 yr.

The initial condition is a temperature at all depths, by assumption, of T_o. The boundary condition on the surface, x of zero, is a seasonal change of temperature, ΔT from summer to winter, with t in years.

$$\frac{\partial T}{\partial t} = \frac{a \partial^2 T}{\partial^2 x}$$
$$T(x,0) = T_o \qquad (4.14)$$
$$T(0,t) = T_o + \Delta T \sin(2\pi t)$$

The printout of the initial dialogue for "Pipe_Freeze" is shown in Figure 4.27. The temperature below ground is tracked for a number of years and the depth for which the temperature just reaches zero is printed which is about 0.76 meters. The surface temperatures are assumed go from $-15°C$ to $35°C$ through the seasonal cycle.

The solution is shown in Eq. (4.15). In this case T_o is the mean surface temperature of $10°C$ and the temperature swing is plus and minus $25°C$. The frequency ω is the seasonal driving frequency.

$$T(x,t) = T_o + \Delta T^* e^{-x\sqrt{\omega/2a}} * \sin\left(\omega t - x\sqrt{\frac{\omega}{2a}}\right) \qquad (4.15)$$

```
>> Pipe_Freeze
   Look at heat tranfer from Earths surface into the ground


Seasonal temperature vs. depth r below ground
>> Pipe_Freeze
   Look at heat tranfer from Earths surface into the ground


Seasonal temperature vs. depth x below ground
 Enter the number of years to track: 2
Minimum Depth to Avoid Freeze = 0.757576
```

Figure 4.27: Printout of the dialogue in "Pipe_Freeze".

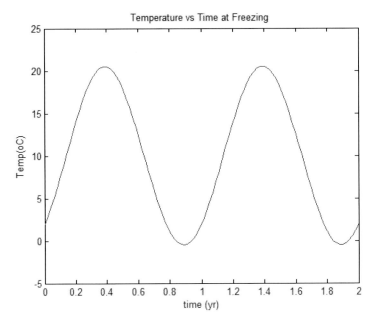

Figure 4.28: Time dependence of T at a depth where the minimum temperature is just at freezing. Note the time lag with respect to surface temperatures at this depth.

The result for a minimum temperature of just freezing is shown in Figure 4.28. That depth is evaluated using the MATLAB utility min for argument (abs(T)). Note the time lag with respect to surface temperature as the heat propagates into the ground.

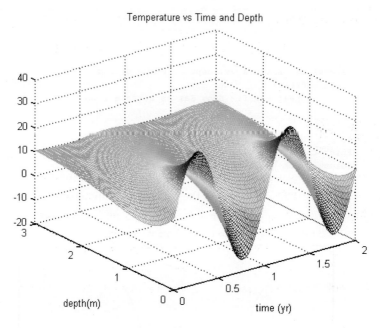

Figure 4.29: Mesh plot of the temperature as a function of time and depth. The seasonal temperature peak propagates in both depth and time, which causes a time lag with respect to the surface. At large depths the mean surface temperature is approached.

The contour for temperature as a function of time and depth appears in Figure 4.29. The temperature is followed over a two-year cycle. There is a clear temperature delay with time as the changes in temperature propagate underground.

4.12. Cooling Earth — Interior

The cooling of the Earth as heat diffuses into space is considered in the script "Cooling_Earth_Int". A series solution to the heat equation is evaluated with thirty terms in the series shown in Eq. (4.16). The initial condition is that the Earth's temperature, T_E, is uniform on the surface at r equal to R. The temperature difference between the Earth and space, T_s, is taken to be three hundred degrees absolute at time zero. There is a characteristic time, τ, which depends on the radius of the Earth and the diffusion coefficient. For reasonable values of the coefficient a the time is one hundred and thirty billion years, even neglecting internal heating as is done here

$$T(R,0) = T_E, \quad dT(r,0) = T_E - T_s = 300°\text{K}$$
$$\tau = \frac{1.0}{(\pi a R)^2} \tag{4.16}$$
$$T(r,t) = 2 \sum_n (-1)^{n+1} T_E \left[j_o\left(\frac{n\pi r}{R}\right) \right] e^{-n^2 t/\tau}$$

The solution is a series in radius and time factors. The radial terms are spherical Bessel functions, using the MATLAB utility "besselj", of argument $(n\pi r/R)$ as is expected in the spherical geometry. The temporal terms are exponentials with arguments $(n^2 t/\tau)$, so that the series converges rapidly in time.

The results for thirty terms in the series are plotted below. In Figure 4.30 the last frame of a movie shows the temperature in the interior. The surface has cooled to zero absolute but the central

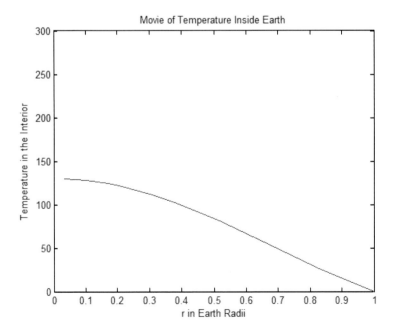

Figure 4.30: Temperature distribution, absolute scale, inside the Earth after 200 billion years.

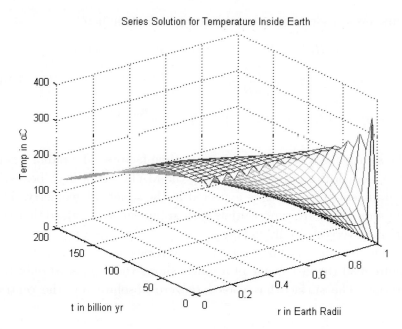

Figure 4.31: Temperature distribution inside the Earth as a function of radius and time. Initially the Earth is at a uniform temperature. As time goes on, the surface temperature falls rapidly, while the core temperature is maintained for a long time.

interior is still above one hundred degrees absolute. A mesh plot of the temperature as a function of time and depth appears in Figure 4.31, showing rapid surface cooling.

4.13. Cooling Earth — Exterior

The exterior temperature as the Earth cools is formally very similar to the previously quoted interior solution. The Earth is taken to be a uniform sphere and the time scale is the same as before. The initial conditions are that the temperature at the surface is T_E at time zero and that at late times the temperature at radius R approaches T_s. The solution is again a series of spherical Bessel function terms spatially, and exponential decaying factors in time, just as before.

The results for the exterior, with the same parameter values as taken for the interior solution shown previously appear in Figure 4.32.

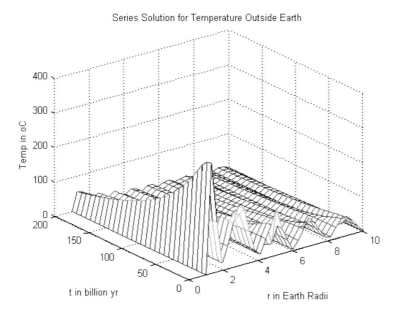

Figure 4.32: Surface of exterior temperatures for a cooling earth. The radii are greater than the earth's radius R. The characteristic time scale is 130 billion years as before.

The mesh plot is for a series with thirty terms and with radial values greater than R. At radius of R there is a smooth decrease in temperature as there was in the interior case. At values of r greater than R, the temperature falls, with structure due to the Bessel functions. Only the very first few terms are important because the exponentially decreasing time factors damp out the higher order terms.

4.14. Heat and pde

The script "Heat_Diffuse_pde" uses the MATLAB one-dimensional partial differential equation solver, "pdepe" to address the problem of tracking the temperature over an x region, $(-L, L)$ in reaction to a pulse of heat initially in the x region $(-ab, ab)$. There is a menu driven choice of different functions for the initial pulse as shown in Figure 4.33. A movie is shown of the time development of the temperature. The script supplies functions defining the heat equation (source less), the initial conditions and the boundary conditions.

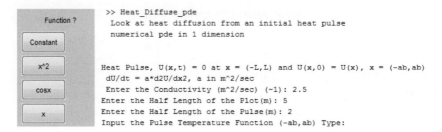

Figure 4.33: Menu of initial temperature functions (left). Dialogue for the "Heat_Diffuse_pde" script for a specific set of user choices (right).

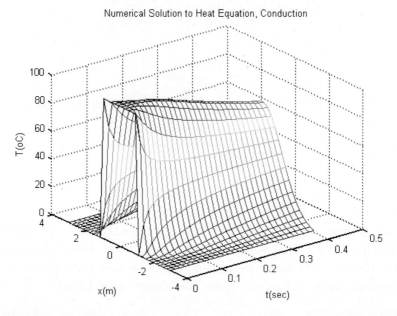

Figure 4.34: Temperature as a function of x and t for an initially uniform distribution.

The temperature response as a function of x and t to a constant temperature initial pulse is shown in Figure 4.34, while the response to an initial temperature which is quadratic in x is displayed in Figure 4.35. In both cases distinct shapes in position for the temperature distribution quite rapidly smooth out and the overall temperature scale dies off in time. All the possible shapes available to the user have the same general behavior.

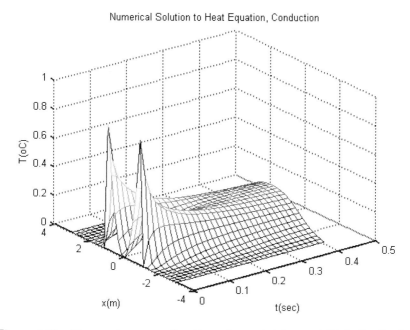

Figure 4.35: Temperature as a function of x and t for an initial distribution quadratic in x.

4.15. Heat Diffusion

The script "Heat_Cond2" uses the MATLAB solver for a partial differential equation much as in the "Heat_Diffuse_pde" script and describing a bar of conductor of length L. However the initial conditions in this case are that the bar temperature is zero, while the boundary conditions at x near L is that the end heats very rapidly to T_L, while the temperature at x of zero is held at zero. The problem approximates the heating of a rod as heat flows into the rod which has fixed temperatures at both ends of the rod.

The printout for "Heat_Cond2" appears in Figure 4.36, where the user chooses the length of the copper rod and T_L. The T_L heating time is small with respect to the time to relax to equilibrium. The characteristic relaxation time is $1/a(\pi/L)^2$, in this case 250,000 seconds. A movie is provided of the time development of the temperature. The printout defines the initial conditions, IC, and boundary conditions, BC.

A mesh plot for the parameter choices made in the dialogue, Figure 4.36 as a function of x and t is shown below in Figure 4.37. As can be seen the ends of the rod are fixed at five hundred degrees and zero degrees. As time increases, the temperature along the

```
>> Heat_Cond2
   solve heat equation for conduction - linear rod
   function Heat_Cond2 - heat at x = L quickly


   Cu rod, length - L
   IC, u(x,0) = 0
   BC, u(0,t) = 0, u(L,t) = TL(1-exp(-t/tau));
   Enter Length of Rod (m): 5
   typical conduction time = 250000 (sec)
   Enter temperature at x = L, long times = TL (oC) : 500
   Enter Heating Time at x = L (sec) : 10
```

Figure 4.36: Printout of the dialogue for a set of specific parameter choices for the script "Heat_Cond2".

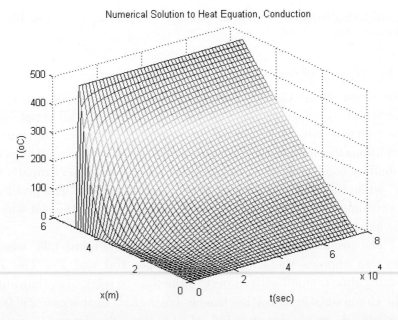

Figure 4.37: Mesh plot of the temperature as a function of position along the rod and time since initial heating of one end of the rod. The temperature along the rod near $x = 0$ increases until a linear temperature dependence on x is obtained.

rod approaches the equilibrium solution of a linear temperature distribution.

4.16. Heat — Initial

In this script, "Heat_IC", the heat equation is solved for the case of a symbolic, user supplied, initial temperature distribution in x from x of zero to x of L. The solution, $T(x,t)$ is by separation of variables and series solution. The assumed boundary conditions are that the partial derivative of T as a function of x for all t are zero at x of zero and L. No temperature flows out of the defined region. The series has Cartesian Fourier spatial terms and exponential temporal terms as seen in Eq. (4.17). The characteristic time for the problem is $(L^2/a\pi^2)$. The A coefficients are solved for symbolically using MATLAB "int".

$$T(x,t) = \sum_i A_i e^{-a(i\pi/L)^2 t} \cos\left(\frac{i\pi x}{L}\right)$$

$$A_o = \left(\frac{1}{L}\right) \int_0^L T_o(x)dx, \quad A_i = \left(\frac{2}{L}\right) \int_0^L T_o(x) \cos\left(\frac{i\pi x}{L}\right) dx$$

(4.17)

A movie of the x dependence of the temperature as a function of t is provided by the script. Printout for a specific choice of initial condition is shown in Figure 4.38. The user chooses the length of the rod and the initial symbolic temperature distribution. The number of terms of the series is also a user choice so that the convergence

```
>> Heat_IC
    Solve Heat Eq. for BC - flow = 0 at boundary and IC temp distribution

Heat Eq. for 0 flow out of end points and defined initial temp distribution
Length of Rod in x is L = 1
dT/dx = 0 at x = 0 and = L
IC, T(x,0) - To(x)
: abs(x-1/2)^2
Enter the Number of terms in the Series: 5
tau for relaxation = 1/(pi^2*a) = 1.01321
```

Figure 4.38: Printout provided by "Heat_IC". The initial temperature distribution is symbolic and provided by the user as a function of x.

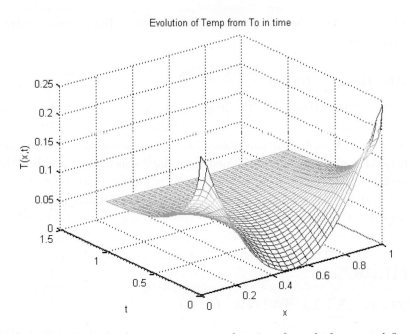

Figure 4.39: Mesh plot for temperature as a function of x and t for a user defined initial distribution, $T(x,0) = |x - 1/2|^2$. Note that the equilibrium temperature is non-zero since the initial temperature has a non-zero average value and no heat flows out of the rod.

of the series can be explored. The mesh plot of temperature as a function of x and t is shown in Figure 4.39. The initial distribution in x is smoothed and reduced as time increases. However, the solution at large t is a constant because the initial distribution had a non-zero average value, and heat does not flow out of the rod by construction.

Chapter 5

Waves

"If you want to find the secrets of the universe, think in terms of energy, frequency and vibration."
— **Nikola Tesla**

"Sit in reverie and watch the changing color of the waves that break upon the idle seashore of the mind."
— **Henry Wadsworth Longfellow**

5.1. Longitudinal Slinky

It is useful to consider longitudinal waves in a continuous medium first as oscillations in a finite discrete set of springs and then go to the continuum limit. Such waves are well known in the "Slinky" toy. The problem is approached in a simplified case of seven springs connecting eight mass points in the script "Slinky" which solves for the motion numerically using the MATLAB script "ode45".

The motion of a mass point with coordinate δ_i due to nearest neighbor displacements is governed by the equation:

$$\left(\frac{m}{a}\right) d^2\delta_i \, dt^2 = \left(\frac{k}{a}\right) \left[+(\delta_{i+1} - \delta_i) - (\delta_i - \delta_{i-1})\right] \quad (5.1)$$

where m is the point mass, k the spring constant and a is the equilibrium separation of the mass points. It is amusing to notice that the grid method of solving partial differential equations uses the spatial part of the second order partial spatial derivative. In the present case the aim is to go to the continuum limit, but actual numerical solutions go the other way on a discrete grid.

The longitudinal motion of the mass points is displayed as a "movie". Only the fourth mass point is displaced initially and all initial velocities are zero. The equilibrium separation, a, is defined to be equal to one. The user supplies the ratio k/m and that defines the characteristic time scale, $1/\sqrt{k/m}$. The motion of the springs

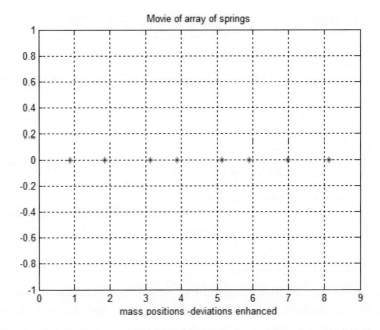

Figure 5.1: Last frame of a movie with initial displacement of the fourth mass point by $0.01a$.

is followed for five such times. The user also supplies an initial displacement of spring #4. In the example shown it is 0.01. The last frame of that specific movie is shown in Figure 5.1 where the displacements are magnified by a factor fifty in order to make them more visible. The movie makes the outgoing left and right waves very clear.

Going to a continuum limit, the coordinates are the fractional displacement, $\eta \rightarrow (\delta_{i+1} - \delta_i)/a$. The Young's modulus $Y \rightarrow ka$ and the density $\rho \rightarrow m/a$. The wave equation in this case follows plausibly from Eq. (5.1) for the discrete case.

$$\frac{\rho \, d^2\eta}{dt^2} - \frac{Y d^2\eta}{dx^2} = 0 \qquad (5.2)$$

As an example, the speed of sound in the atmosphere can be estimated using the wave equation. Take the ratio of specific heats, appropriate to adiabatic motion where there is not sufficient time to carry away the heat generated, $\gamma = C_P/C_V = 5/3$. The adiabatic approximation assumes $dP/P = -\gamma(dV/V)$ which relates the fractional change in pressure to the fractional change in volume. The atmosphere at STP has a pressure of 0.101 MPa and a density of 1.29 kg/m³. The velocity, $v = \sqrt{\gamma P/\rho}$, can be read off from the wave equation, Eq. (5.2) since the adiabatic Young's modulus is γP with units of pressure. Numerically the result for the speed of sound is 360 m/sec.

Plots of the displacements of individual mass points with time are shown in Figures 5.2 and 5.3. A longitudinal wave has clearly passed from one mass point to the next. The wave arrives first at mass #2, elongates it and then compresses it while the elongation appears then at #1, followed by compression. The wave velocity can be taken for the two figures. In this specific case the wave moves by a distance a of one in about one characteristic time unit.

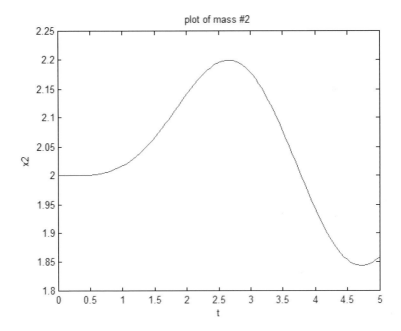

Figure 5.2: Plot of the position of mass point #2. Initially it does not move and then the longitudinal wave reaches it and it is elongated, then compressed. Note that the displacement is greatly magnified.

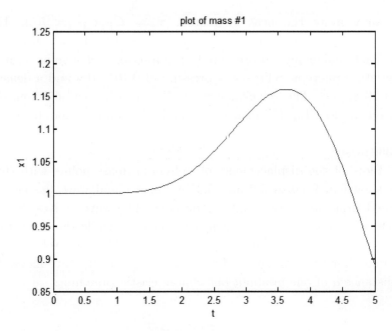

Figure 5.3: Plot of the position of mass point #1. Initially it does not move and then the longitudinal wave reaches the mass, after the wave has hit mass #2. There is first elongation followed by compression. Note that the displacement is greatly magnified.

5.2. Transverse Slinky

The transverse oscillation modes for a short "slinky" are studied in the script "Transverse_Slinky". This script uses the MATLAB function "eig" to find the eigenvalues and eigenfunctions for $N = 2, 3$ mass points connected by springs with the end mass springs fixed to rigid walls and with $k/m = 1$ for the springs. The eigenvalues for an N of two and three are shown in Figure 5.4, while the eigenfunctions are graphically displayed in Figure 5.5 for $N = 2$ and Figure 5.6 for $N = 3$.

In the case of N equal to two, the eigenfunctions are transverse motion in the same or opposite directions. In the case of N equal to three, the central spring is stationary for the lowest state and the two outer springs have opposite displacements. For the next lowest eigenvalue, the two outer springs have the same displacement,

Figure 5.4: Eigenvalues, D, for $N = 2$ (left) and $N = 3$ (right) mass points.

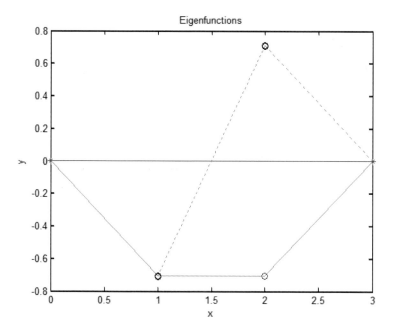

Figure 5.5: Eigenfunctions for $N = 2$ mass points. The end points are fixed by springs attached to rigid walls. The lowest eigenvalue is shown as a dotted line.

while the middle mass has the opposite displacement. With the largest eigenvalue, all three mass points have the same sense of displacement.

5.3. SR — Doppler

The Doppler effect is well known in everyday life. However, there are additional changes to that picture as the source velocity approaches that of light. The ratio of the frequencies in two inertial frames

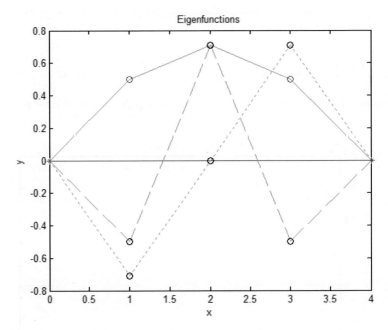

Figure 5.6: Eigenfunctions for $N = 3$ mass points. The end points are fixed by springs attached to rigid walls. The eigenfunction for the lowest eigenvalue is dashed, followed by dashed and solid for the higher eigenvalues.

moving with a relative velocity $v = \beta/c$ is:

$$\frac{\omega_*}{\omega} = \frac{1}{\gamma}(1 + \beta \cos(\theta_*)) \tag{5.3}$$

In the classical theory there is no shift in frequency at 90 degrees to the direction of motion. In the relativistic case there is a shift, as illustrated in Figure 5.7 which is output of the script "SR_Doppler". This transverse Doppler shift is due to the time dilation factor between the two frames. The longitudinal shift is $\omega_*/\omega = \sqrt{(1-\beta)/(1+\beta)}$, while the transverse shift is $\omega_*/\omega = 1/\gamma$. The emission angles of the two frames are related by, $\tan(\theta_*) = \sin\theta/\gamma(\cos\theta - \beta)$. The result is easily derived using the fact that the wave vector k and the frequency, ω/c are the components of a four-vector which transforms using the Lorentz transformation.

A contour plot of the ratio of the frequencies as a function of β and the emission angle of the light in the $*$ frame is displayed

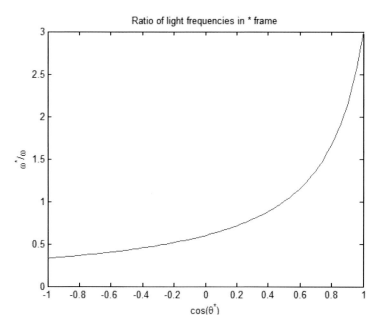

Figure 5.7: Ratio of frequencies of the light observed in 2 inertial frames moving with a velocity factor $\gamma = 1/\sqrt{(1-\beta)*(1+\beta)}$, for the case where the factor is 2 as a function of the angle of the light with respect to the axis of relative motion.

in Figure 5.8. If the source is receding, the light is red shifted to lower frequencies while a source moving toward the frame is blue shifted. The velocity is a signed quantity as is the emission angle since emission can be toward the direction of relative motion or away from it. The emission is symmetric if the sign of the velocity and the direction of the light emitted are both changed.

5.4. Step Response

A damped oscillator has a complicated response to a step driver which starts at time zero. The differential equation has a damping term, a, a natural frequency in the undamped case, b, and a driving term c.

$$\frac{d^2y}{dt^2} + 2a\left(\frac{dy}{dt}\right) + b^2 y = c, \quad t > 0 \tag{5.4}$$

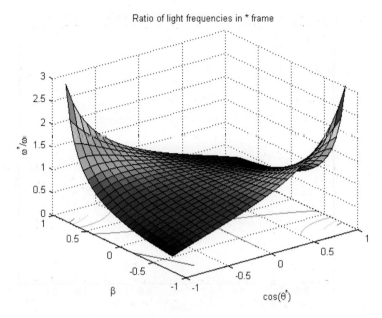

Figure 5.8: Contour of the ratio of frequencies in 2 inertial frames for different relative velocities and for different emission angles in the ∗ frame.

The problem is solved symbolically using the MATLAB utility "dsolve" in the script "Step_Transient". The solution is reproduced in Figure 5.9. At long times after the transients die out the amplitude is a constant. Examining the differential equation, in this case $y \to c/b^2$ is expected.

The amplitude for a specific, user defined, example is shown in Figure 5.10. There is initial transient behavior which is damped out on a time scale set by the scale of the natural frequency $\tau \sim 1/\omega \sim 1/b$. The solution asymptotically approaches a constant value, as expected. A constant driving term, c, is assumed, but other functional forms could be tried, such as a harmonic driver.

5.5. Pulsating Sphere

An ideal, incompressible fluid will propagate waves driven by a harmonically varying object, for example water is a good transmitter of sound waves. A specific case is considered in the script "Pulsating_Sphere". The radial velocity of the sphere is harmonic

```
 c     c exp(-t (a + #1))  (a - #1)     c exp(-t (a - #1))  (a + #1)
 -- +  ------------------------------ - ------------------------------
  2                 2                                2
  b               2 b   #1                         2 b   #1

where

                        1/2
 #1 == ((a + b) (a - b))
```

Figure 5.9: Printout of the script "Step_Transient" showing the symbolic solution, y, when the initial position and velocity of the oscillator are zero.

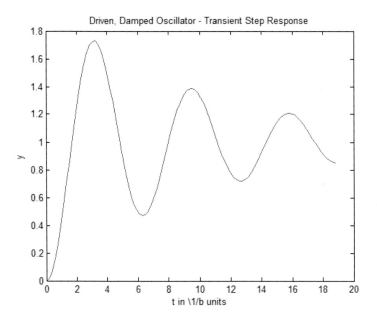

Figure 5.10: Amplitude for the damped, driven oscillator in the specific case of $a = 0.2, b = c = 1$.

with a circular frequency ω. The resulting velocity of the perfect fluid as the wave propagates and dissipates is:

$$v(r) = v_o \left(\frac{a}{r}\right)^2 \frac{[c^2 + ar\omega^2)\cos\phi + c\omega(r-a)\sin\phi]}{[c^2 + (a\omega)^2]}$$

$$\phi = \frac{\omega(r-a)}{(c-\omega t)}$$

(5.5)

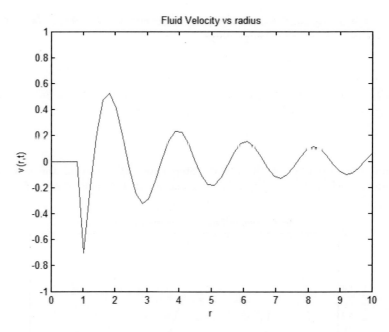

Figure 5.11: Radial velocity of the wave in the fluid for a driving frequency of 3. The sphere is of radius 1 and the velocity is zero for $r < 1$.

The amplitude of pulsation, v_o, is taken to be one as is the radius a of the sphere and the sound velocity in the fluid, c. A movie of the radial dependence of the velocity as a function of time is plotted by the script. A frame of that movie in the case of a driving frequency of ω equal to three, a particular user choice, appears in Figure 5.11. Note the damping with radius as the wave spreads as indicated by the inverse square of the radius behavior shown in Eq. (5.5). The frequency of the wave is the driving frequency in the case of a perfect fluid, without viscosity or other dissipation.

5.6. Rectangular Drum

The rectangular drum is constrained to have a vanishing displacement on the boundaries at x of zero and a and in y at zero and b. The frequencies thus are set for the modes n and m in the spatial solutions which are printed out in the dialogue with the script. Temporally, the choice was made that the tension and density of the drum were both defined to be one, so the drum wave velocity is one from the wave equation.... The user defines the extent in b and the modes m and n. The dialogue for the script "Rectangular_Drum" appears in Figure 5.12.

The plotted output is a movie of the motion of the drum in time displayed at the displacement z as a function of x and y as time progresses. A frame of that movie is shown in Figure 5.13. The user

```
Rectangular Drum of x Width a = 1
  Spatial: sin(m*pi*x/a)sin(n*pi*y/b)
  Temporal: cos(2*pi*vn,m*t), vm,n = 1/2*sqrt(T/rho)*
  Temporal: sqrt((m/a)^2+(n/b)^2)
 Enter the x mode m: 2
 Enter the y mode n: 3
 Enter the y Extent - a = 1, b =: 2
```

Figure 5.12: Printout of an example of the dialogue from the script "Rectangular Drum".

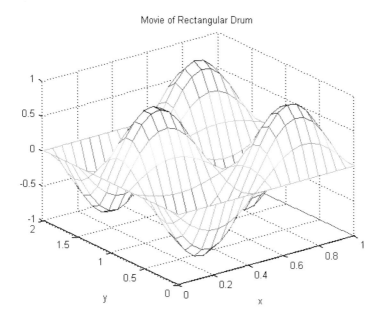

Figure 5.13: Solutions to the 2-dimensional wave equation in rectangular coordinates and with rectangular boundary conditions imposed. The choices made are: size is 1 in x and 2 in y. The modes are 2 in x and 3 in y. There are 2 extreme values in x and three in y.

is encouraged to vary b, m and n in order to see the changes induced in the drum by watching the resulting movie.

5.7. Piano

The script "Piano" approximates the motion of a piano string which is struck by an impulsive force, the piano hammer, at location ax_o along the string. The string is held at $x = 0$ and $x = L$ where L is taken to be one in the script. The velocity of waves on the string is c, taken here to be equal to one. The solution for the displacement of the string, $y(x,t)$ depends on the location of the hammer and the time t. The impulse is approximated by a function which is large only for times $< \pi/q$ where q is chosen to be very large. The time t is zero at the end of the impulse and the string position, y, is plotted only for t greater than zero. The series solution for the piano string displacement appears in Eq. (5.6). The boundary conditions are satisfied by the sin terms in the series, while the initial quasi-impulse is set by the cos term.

$$y(x,t) \sim \sum_1^\infty \frac{\sin\left(\frac{n\pi x_o}{L}\right)\sin\left(\frac{n\pi x}{L}\right)\sin\left(\frac{n\pi c t'}{L}\right)\cos\left(\frac{\pi^2 c}{Lq}\right)}{\{n[(n\pi c)^2 - (qL)^2]\}} \qquad (5.6)$$

A movie is produced after the initial impulse position is chosen and the number of terms in the series is defined. The last movie frame in the cases of $x_o = 0.2$ and ten and fifty terms respectively are plotted in Figures 5.14 and 5.15. Note that the waveform is much sharper in the latter case since the Fourier series is more accurate because of the increased number of terms. No dissipation has been assumed in this script.

The movie for the time development of the string exhibits the travelling wave behavior both to increasing x and decreasing z and the reflections off the ends of the string where it is held to zero displacement.

5.8. Aperture Diffraction Off Axis

Diffraction by an aperture has two distinct modalities. For wave vector k times aperture size a, when ka is much greater than one

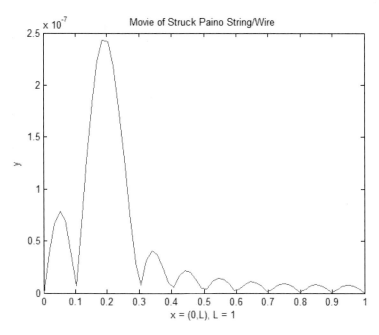

Figure 5.14: Initial amplitude of the piano string using 10 terms in the series representation and initial $x = 0.2$.

the regime of geometric optics is operative where light appears to go in straight lines. For ka small with respect to one the diffraction of light by any aperture becomes very evident.

For a circular aperture of radius a and angle of incidence with respect to the normal of plate containing the aperture of angle ψ, the angular distribution of the power, $dP/d\Omega$ downstream of the aperture is:

$$\delta = \sqrt{\sin^2 \theta + \sin^2 \psi - 2 \sin \theta \sin \psi \cos \phi}$$
$$\frac{dP}{d\Omega} = (ka)^2 (\cos^2 \phi + \cos^2 \theta \sin^2 \phi) \left| \frac{2 J_1(ka\delta)}{ka\delta} \right|^2$$

(5.7)

The azimuthal angle, φ, is the angle of the electric field of the incident wave with respect to the outgoing wave vector x axis and is taken to be zero. The angle θ is the spherical polar angle of the wave vector downstream of the aperture. The MATLAB utility "besselj" is used to evaluate the numerical values of the radiated power.

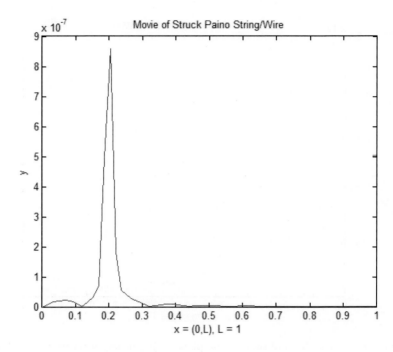

Figure 5.15: Initial amplitude of the piano string using 50 terms in the series representation and initial $x = 0.2$.

The user can check on expectations in regards to geometrical optics by entering in the script "Diffract_Angle" a zero angle and a large value, say 20, for ka. There is an azimuthally independent small angle peak within the diffractive cone of angle approximately $1/ka$. The situation for $ka = 1$ and normal incidence is shown in Figure 5.16. There is, indeed, a forward diffractive peak, but also azimuthal dependence. The azimuth of the incident electric field is important and the power peaks at that azimuth.

For $ka = 1$ and an incident angle of forty-five degree, the pattern of radiated power is shifted toward larger angles with respect to normal incidence (Figure 5.17). There remains a forward peak, but the overall pattern is shifted toward larger polar angles with a modified azimuthal dependence. The user can select both ka and the incident angle and thus explore the parameter space.

5. Waves 153

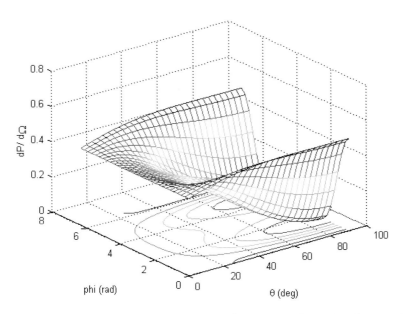

Figure 5.16: Radiated power for the case of $ka = 1$ and zero incident angle. There is a forward diffractive peak in θ and a dependence on φ.

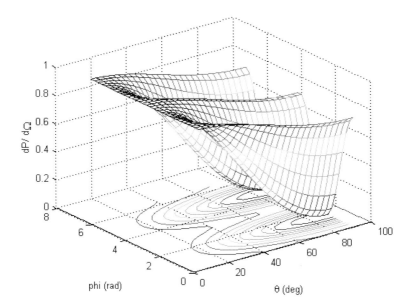

Figure 5.17: Radiated power for the case of $ka = 1$ and a 45 degree incident angle. The peak is still forward but wider angles are now more probable.

5.9. Antenna

The power radiated by a center fed linear antenna is displayed using the script "Antenna". The antenna length is L; the antenna resistance is R_o; and the crest current at the antenna midpoint is I_o. The length L is in units of the driving current wavelength. Printout of the script appears in Figure 5.18. The printout displays the angular distribution of the radiated power for a user defined choice of wavelength.

The angular distribution of the radiated power in the case where $kL = 0.1$ is shown in Figure 5.19. The pattern goes as the square of the sin function as expected for dipole radiation. The total radiated power is shown in Figure 5.20 as a function of kL. A closed form is not available in this case, so that the MATLAB numerical utility "quad" is used to integrate the angular distribution. The power increases rapidly with kL. For $kL < 1$ the approximate results for the angular distribution and the total power are quoted in Eq. (5.8). The total radiated power goes as the square of the antenna length, with a scale set by the wavelength of the radiated waves.

$$\frac{dP}{d\Omega} \sim \frac{\cos^2\left(\frac{\pi}{2\cos\theta}\right)}{\sin^2\theta}$$

$$P \sim \pi R_o I_o^2 \left(\frac{L}{\lambda}\right)^2 \tag{5.8}$$

```
>> Antenna
   Program to look at radiated power of a center fed antenna

   Antenna Length = L, Io = crest current at center

        2    /     / L \      / L cos(t) \ \2
   Io  Ro | cos| - | - cos| --------- | |
             \   \ 2 /      \     2     / /
   -------------------------------------
                  / L \2       2
          4 pi sin| - |  sin(t)
                  \ 2 /
```

Figure 5.18: Printout of the script "Antenna".

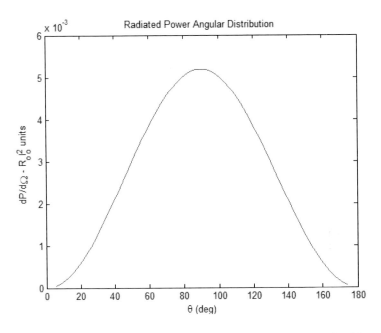

Figure 5.19: Angular distribution of the radiated antenna power for $kL = 0.1$.

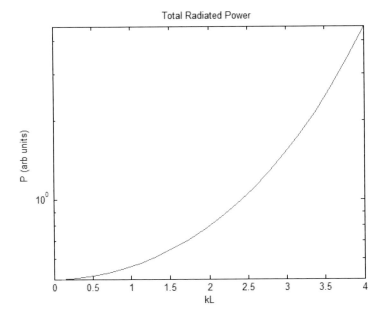

Figure 5.20: Radiated antenna power as a function of kL. The semilog plot in y indicates the strong L dependence.

5.10. Antenna Array

Simple dipole antennas are quite isotropic in their radiation pattern as seen above for small values of kL. The use of two antennas with controlled phase between them was already explored in Section 3.17. In that case, substantial directionality could be obtained. However, often one wishes to more tightly "beam" a radio or other signal to a specific angular range. This can be accomplished by having different waves from different parts of an antenna array interfere with one another. In the specific case of an array of N quarter wave, $\lambda/4$, antennas separated by spacing a and with relative phases φ_o, the resulting intensity, I, is shown in Eq. (5.9).

The full array pattern can be thought of as N slit diffraction due to the contributions of the N line sources. There is a maximum of I at the sin of the angle equal to $2(1/N - \varphi_o/\pi)$. The pattern depends on N, the total number of antennas, and the relative phase of the antennas, φ_o.

$$\frac{\Phi}{N} = \frac{\frac{(\pi \sin\theta)}{4+\phi_o}}{2}$$

$$\Phi_2 = \frac{(\pi \sin\theta)}{4+\phi_o} \qquad (5.9)$$

$$I = \left(\frac{\sin\Phi}{\sin\Phi_2}\right)^2$$

The script "Antenna_Array" allows the user to choose both the number of members of the array, and the phase of the elements. Radiated power patterns are shown in Figures 5.21 and 5.22. It is obvious that with a large number of elements in the array, very tightly beamed power can be obtained. By varying N, the user can watch the change in the power angular distribution. Clearly, this beaming technique has important practical implications.

5.11. Lissajous

A particle may be confined in two dimensions with different confining potentials in the two directions. In that case there can be simple

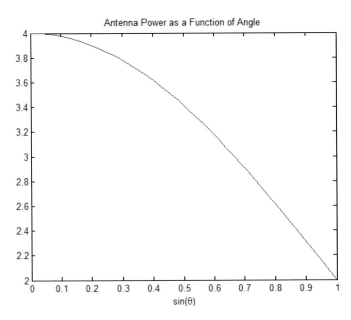

Figure 5.21: Intensity for $N = 2$ and phase equal 0. For $N = 1$ the power is emitted isotropically in this approximation.

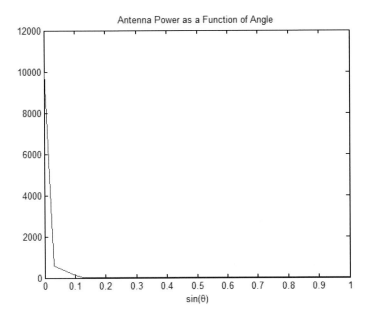

Figure 5.22: Intensity for $N = 100$ and phase equal 0.

harmonic motion in each dimension, but with different amplitudes, phases and circular frequencies. It can then happen that the trajectories in the (x,y) plane can become rather complex. The x and y trajectories differ in frequency, ω_x, ω_y, phase φ and the amplitude ratio which is defined to be a.

$$\begin{aligned} x &= \cos(\omega_x t) \\ y &= a\cos(\omega_y t + \phi) \end{aligned} \qquad (5.10)$$

The solutions are shown in the script "Lissajous" as movies where the user chooses the parameters for the x and y harmonic oscillations. The patterns may be re-entrant, but need not be. A plot of (x,y) for a specific case is the last frame of the movie generated with an amplitude ratio of two, a phase difference of $\pi/2$ and a frequency ratio of four is shown in Figure 5.23. Such plots are familiar to devotees of old science fiction movies where oscilloscopes are driven in (x,y) to make amusing "laboratory" displays.

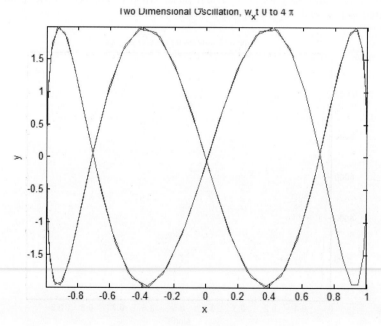

Figure 5.23: Plot of (x,y) for simple harmonic motion in 2 dimensions with differing amplitudes ($2\times$), phases ($\pi/2$) and frequencies ($4\times$).

5.12. Plane Waves — 2d

Two waves propagating on an infinite membrane display some interesting interference effects. A simple case is that of two waves of equal amplitudes, frequencies, ω, and wave vectors of the same magnitude, k, with the only difference being that the angle between the wave vectors is finite. The x axis is chosen as the bisector of the angle between the two wave vectors. The superimposed amplitude is easily shown to be, with $k_x = k\cos(\theta)$, $k_y = k\sin\theta$ and θ the angle between the two waves.

$$\cos(k_y y) e^{i(k_x x - \omega t)} \tag{5.11}$$

The superimposed waves travel in a new, x, direction with a modulated amplitude in the y direction and with an increased wave velocity ω/k_x. The script "Plane_Wave_2d" is used to choose the angle between the two individual waves and the resultant movie is shown, the last frame of which appears in Figure 5.24. The nodal

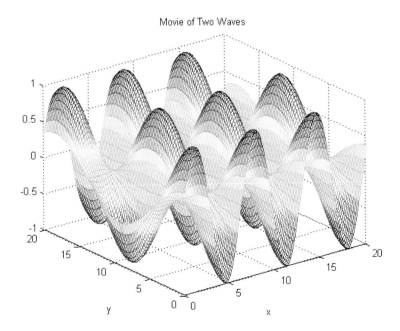

Figure 5.24: Superimposed waves in two dimensions for a 20 degree angle between the 2 constituent waves. There are two nodal lines in y which are evident.

lines for the standing wave which exists in the y direction are:

$$y_n = \frac{\left(\frac{n-1}{2}\right)\pi}{k_y} \qquad (5.12)$$

5.13. Damped, Driven Oscillator

A general second order inhomogeneous differential equation appears in Eq. (5.13). It can be simplified by expressing time in units of the undriven and undamped circular frequency ω_o, or τ. When the natural frequency is defined to be one, there remain three parameters defining the equation, a damping factor b, a driving amplitude C and the ratio of the driving frequency to the natural frequency k.

$$\frac{md^2y}{d^2t} + \frac{ma\,dy}{dt} + by = A\sin(\omega\tau)$$

$$\frac{d^2y}{d^2\tau} + \frac{b\,dy}{d\tau} + y = C\sin(kt) \qquad (5.13)$$

$$\tau = \omega_o t, \quad \omega_o^2 = \frac{b}{m}, \quad k = \frac{\omega}{\omega_o}$$

The equation is explored using the script "Damped_Forced_SHO". The solutions are created at four levels of complexity using the MSATLAB symbolic solutions "dsolve" with initial conditions the y equals one and dy/dt equals zero at t of zero. First the case of no driving force and no damping results in purely harmonic motion. The specific waves associated with Figure 5.25 appear in Figure 5.26. The second condition is for damping but no driving force, while the third has no damping but a driving force. Finally, the general case is considered, with the parameters b, C and k supplied by the user.

The simplest case is just harmonic. The case with damping can be understood by assuming a harmonic solution to Eq. (5.13) and then solving for the frequency ω_b. The frequency is shifted from that for the free and undamped case and acquires an imaginary part illustrating the damping of the wave, Eq. (5.14).

$$\frac{\omega_b}{\omega_o} = \frac{\sqrt{1-\left(\frac{b}{2}\right)^2}+ib}{2} \qquad (5.14)$$

5. Waves

```
Simple Harmonic, No Damping, No Driving

   C2 cos(t) + C3 sin(t)
Simple Harmonic, Damped, No Driving

        /      / b     #1 \ \              /     / b    #1 \ \
    exp| - t | - - --  |  | (b + #1)   exp| - t | - + -- |  | (b - #1)
        \     \ 2     2 / /              \     \ 2    2 / /
    -----------------------------  -  -----------------------------
                 2 #1                             2 #1

  where
                             1/2
        #1 == ((b - 2) (b + 2))
Enter Damping b : 1
Simple Harmonic, No Damping, Driving

                      /   c k        c k   \           / c cos(t (k - 1))    c cos(t (k + 1)) \
     cos(t) + sin(t) | ------- - ------- | - sin(t) | ---------------- + ---------------- |
                      \ 2 k - 2    2 k + 2 /           \    2 k - 2            2 k + 2       /

                     / c sin(t (k - 1))    c sin(t (k + 1)) \
           cos(t) | ---------------- - ---------------- |
                     \    2 k - 2            2 k + 2        /
Enter amplitude c and frequency k as [ , , ]): [5,1.1]
Simple Harmonic, Damped and Driven
Enter damping b, amplitude c and driving frequency k as [ , , ]): [1 5 1.1]
```

Figure 5.25: Results of the dialogue for the script "Damped_Forced_SHO". There are 4 cases; no damping, no driver, damping, no driver, driver no damping and finally damping with a driving force.

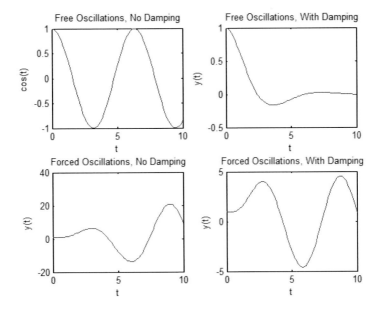

Figure 5.26: Wave forms for the 4 specific cases given in Figure 5.25.

For the driven case, after the transients have died down, the amplitude y follows the driving force amplitude and frequency. In the absence of damping the amplitude can become very large if the driving frequency is near the natural frequency (resonance), while damping limits the amplitude, as seen in Figure 5.26.

Chapter 6

Quantum Mechanics

"Quantum physics thus reveals a basic oneness of the universe."
— **Erwin Schrödinger**

"If quantum mechanics hasn't profoundly shocked you, you haven't understood it yet."
— **Niels Bohr**

6.1. Box

The script "QM_Box" uses the MATLAB symbolic mathematics utilities to evaluate expectation values for a particle in a one-dimensional box in the ground state or $n = 2$, 3 or 4. The box boundaries are $(-a/2, a/2)$. The expectation values for x, momentum p, x^2, and p^2 are calculated symbolically. The energy is evaluated and the uncertainty of the ground state, defined in Eq. (6.1), is explicitly evaluated.

$$\Delta x \Delta p = \sqrt{\langle x^2 \rangle - \langle x \rangle^2} * \sqrt{\langle p^2 \rangle - \langle p \rangle^2} \quad (6.1)$$

Printout from the script appears in Figure 6.1. The energy is printed along with the wave function and normalization. Additional printout appears in Figure 6.2, where the expectation values of position and momentum are computed for the ground state, $n = 1$. Other states are available for the user to choose from. In the ground state the basic Heisenberg uncertainty relation is confirmed:

$$\Delta x \Delta p = 0.57 \hbar \quad (6.2)$$

6.2. Simple Harmonic Oscillator

The simple harmonic oscillator is easily solvable classically. The potential energy is $V = kx^2/2$ or $M\omega^2 x^2/2$. At the turning points

```
>> QM_Box
   Program to look at the solutions in infinite square well
   symbolic calculations


   Box, x = (-a/2,a/2)
   Ground state cos(pi*x/a), first excited sin(2*pi*x/a)
   Quantum # n, = 1 or 2 or 3 or 4: 1
   Quantized energy


        2    2
      pi  hbar
     ---------
         2
      2 a  m
   Wave function and normalization

   psi =

   cos((pi*x)/a)


   norm =

   a/2
```

Figure 6.1: Printout from the script "QM_Box". All quantities are calculated symbolically and then evaluated numerically.

the velocity is zero, so the turning points are at $\pm E/k$, where E is the total energy. The classical velocity as a function of x is then:

$$x_T = \pm\sqrt{\frac{2E}{k}}$$

$$v^2 = \left(\frac{2k}{m}\right)(x_T^2 - x^2)$$

(6.3)

The probability to observe the classical particle goes as $1/v$, which means it is mostly near the turning points. Quantum mechanically, the solutions are Hermite polynomials, which are available using the MATLAB Maple functions, mfun('*H*', *n*, *x*).

```
Mean x and x^2

xav =

0

     2    2
    a  (pi - 6)
    ------------
        2
      12 pi
Mean p and p^2

       0

        2    2
       pi  hbar
    -  ----------
           2
          a
```

dxdp using averages for p, p^2, x and x^2

```
 /        2 \1/2  /   2         \1/2
 |    hbar  |    |  a  (pi - 6)  |
 | -  ----- |    |  ------------ |
 |      2   |    \      12      /
 \     a    /
```

Figure 6.2: Additional printout from the script "QM_Box" for the case of the ground state, $n = 1$. The uncertainty principle is explicitly confirmed.

The solutions are:

$$y = x\sqrt{\frac{m\omega}{\hbar}}, \quad \omega^2 = \frac{k}{m}$$

$$E = \hbar\omega\left(\frac{1}{2} + n\right) \tag{6.4}$$

$$\psi \sim H_n(y)e^{-y^2/2}$$

The correspondence between the classical and quantum mechanical solutions is developed in the script "QM_SHO_Hermite". The user chooses a quantum number n and the two solutions, classical and

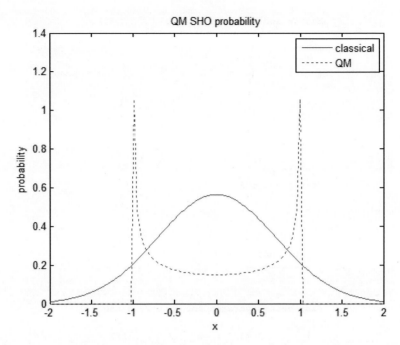

Figure 6.3: Plots of probability for a classical simple harmonic oscillator and for the $n = 0$ quantum case. The wave function is outside the classical turning points at x of $+$ and -1.

quantum, are plotted. Results for $n = 0$, the ground state, and $n = 20$, a highly energetic state, are shown in Figures 6.3 and 6.4, respectively. The solutions are radically different at low values of n but approach one another at high values of n. This behavior is an example of the "correspondence principle" which argues that quantum solutions approach classical ones in the limit of large quantum numbers.

The quantum solution is not limited to be at x values less than the turning points. Indeed looking at the Schrödinger equation and ignoring the energy E, the solution is clearly the Gaussian appearing in Eq. (6.4), $-\hbar^2 \, d^2\psi/dx^2 \sim 2m(kx^2)\psi/2$. The Hermite polynomials are corrections to the basic Gaussian behavior of the wave functions.

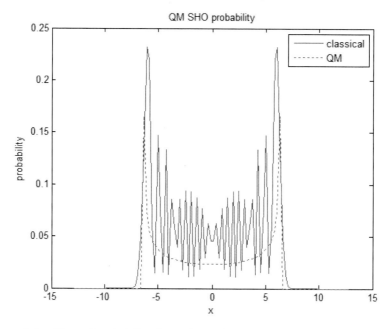

Figure 6.4: Plots of probability for a classical simple harmonic oscillator and for the $n = 20$ quantum case.

6.3. Bound States — 3d Well

The bound state of a three-dimensional square well of radial size a can be solved, but not in closed form in analogy to the one-dimensional square well of finite depth. The potential is equal to V_o for r less than a and zero otherwise. The solutions are spherical Bessel functions inside and Hankle functions outside the well which can be accessed using the MATLAB functions "besselj" and "besselh". For the ground state the radial wave function R, is not used but rather the function u:

$$R = \frac{u}{r}$$

$$u_o(r) = A\sin(kr), \quad k = \hbar\sqrt{2m(E+V_o)} \quad (6.5)$$

$$u_o(r) = Be^{-Kr}, \quad K = \hbar\sqrt{2m(-E_o)}$$

The solution for the ground state is found using the script QM_Square_3d. The script applies the MATLAB utility "fminsearch" to evaluate the energy E, less than zero for a bound state, that satisfies the condition that arises when requiring the wave function and derivative to match at the boundary at $r = a$. The starting value given to "fminsearch" is taken to be the solution for an infinite square well

$$k\left(\frac{1}{\tan(ka) - ka}\right) + K\left(\frac{1+1}{Ka}\right) = 0 \qquad (6.6)$$

Printout from the script is shown in Figure 6.5. Several solutions for a value of a of 1 A are shown. For a well depth of 10 eV the state is barely bound, while for a 24 eV well depth the binding energy is 6.46 eV.

```
>> QM_Square_3d
  Program to look at ground state solutions for a 3-d square well, depth Vo

General solutions are jv(x) and hl(x) - focus on S wave, l = 0
 In general use MATLAB besselj and besselh
Well depth Vo, r < a
Well Depth (eV) -(10,25) : 10
Well radius (A) ~ 1 : 1
 Ground state energy (eV) = -0.00730888
Well depth Vo, r < a
Well Depth (eV) -(10,25) : 12
Well radius (A) ~ 1 : 1
 Ground state energy (eV) = -0.298249
Well depth Vo, r < a
Well Depth (eV) -(10,25) : 24
Well radius (A) ~ 1 : 1
 Ground state energy (eV) = -6.46641
```

Figure 6.5: Printout for some specific choices well depth and extent for the ground state.

The radial wave function rR for the barely bound case is shown in Figure 6.6. The wave function falloff outside the well is very slow since the barely bound state can easily "leak" outside the well. More deeply bound states are more localized to be in or near the well.

To make the comparison concrete, the deeply bound example of a one Angstrom well 24 eV deep is shown in Figure 6.7. In this case

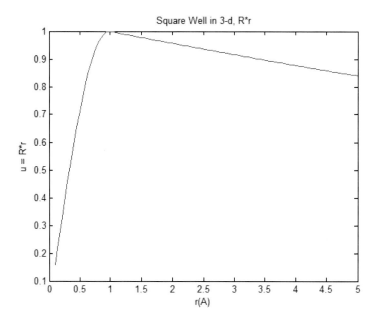

Figure 6.6: Wavefunction for the ground state of a 3d square well when the state is barely bound.

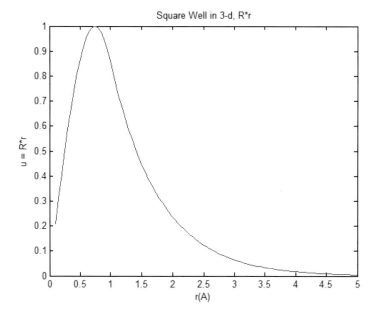

Figure 6.7: Wavefunction for the ground state of a 3d square well when the state is bound by 6.47 eV.

the falloff of the wavefunction outside the well is much faster than in the previous plot, as expected.

6.4. Identical Particles

In quantum mechanics identical particles obey restrictions on the wave functions describing their states. A simple example is to consider two particles confined to a "box" in one dimension. The solutions are sin and cos functions which vanish at the boundaries of the box. Independent particles would be described by product wave functions since the joint probability in such a case is simply the product of the individual probabilities.

However, in quantum mechanics identical particle wave functions must be either symmetric or antisymmetric under the interchange of a pair of particles.

$$\psi_{s,a} = \frac{1}{\sqrt{2}}[\psi_\alpha(1)\psi_\beta(2) \pm \psi_\alpha(2)\psi_\beta(1)] \qquad (6.7)$$

The Greek subscripts indicate the state and the labels 1 and 2 identify the particle in that state. There are two possibilities; bosons have the plus sign for symmetric wave functions under particle interchange while fermions are anti-symmetric.

The wave function contours for several cases are explored using the script "QM_Identical". The user chooses the states to be explored for the box and a plot is made of the resulting wave function contour. The contour for both particles in the ground state is not shown since it vanishes for the antisymmetric case which is the Fermi exclusion principle. For the $n = 1$ and $n = 2$ states of the box, the product contour for classical considerations appears in Figure 6.8. In this case particle 1 and particle 2 are uncorrelated. In contrast the contour for the antisymmetric case is shown in Figure 6.9. In this case there is a sort of "repulsion" when the coordinates are equal. Finally, for the symmetric case, $n = 1$ and $n = 3$ states are plotted in Figure 6.10. Clearly the symmetries imposed on the states of quantum identical particles have profound implications for the physics of a system of such particles.

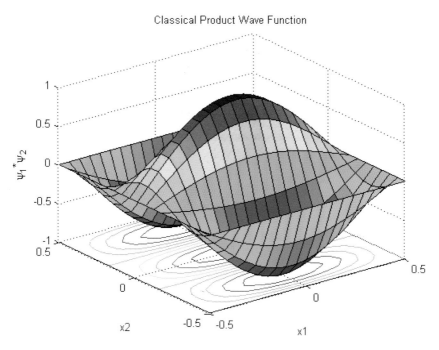

Figure 6.8: Contour of the system wave function for $n = 1$ and $n = 2$ for a product wave function. The axes are uncorrelated.

6.5. Stark Effect

The Stark effect occurs when an external electric field is applied to an atom. In the case of the hydrogen atom in the ground state the first order effect in perturbation theory vanishes. That is because the ground state has no angular momentum while the perturbation defines a vector direction in space, so that the matrix element evaluated in the ground state vanishes. The atomic ground state does however have an induced dipole moment in second order where matrix elements between the ground state and other states contribute.

The effect is first order for degenerate states, the lowest energy being the $n = 2$ hydrogenic state. The perturbation mixes the states with $l = 0$ and $l = 1$ because the operator has vector properties.

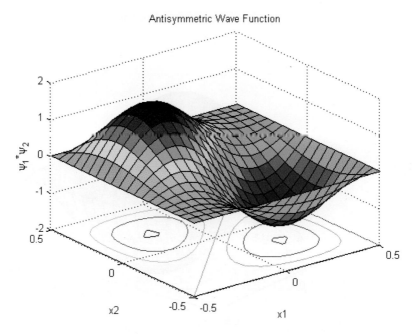

Figure 6.9: Contour of the system wave function for $n = 1$ and $n = 2$ for an antisymmetric wave function. The wave function is highly correlated and is the modulus is small for equal coordinate values.

Printout for this case appears in Figure 6.11. The mixing matrix of the perturbation is A in the printout. The MATLAB utility "eig" is used to find the eigenfunctions V and eigenvalues with the diagonal matrix D of the perturbation.

More printout of the script appears in Figure 6.12. The MATLAB solutions for the eigenvectors, V, and eigenvalues, D, are shown.

A plot of the eigenvectors is shown in Figure 6.13. There is an isotropic part with $l = 0$ and a part with angular dependence. The two perturbed solutions are not symmetric about ninety degrees as the unperturbed eigenfunctions are. That asymmetry means that the electrons have a centroid away from the origin and the displaced charge means that there is an induced dipole moment which is first order in this case. The alignment of the electrons can be either parallel or antiparallel to the external electric field.

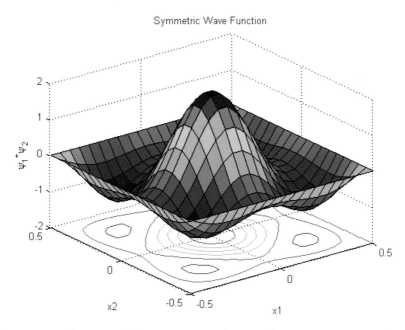

Figure 6.10: Contour of the system wave function for $n = 1$ and $n = 3$ for a symmetric wave function. The wave function is correlated, being large when the coordinates are similar.

6.6. Square Well Scattering — 3d

The Born approximation to a scattering amplitude consists in approximating the incoming and outgoing wave functions by plane waves. In that case, the scattering amplitude, f, is essentially the Fourier transform of the interaction potential with respect to the magnitude of the momentum transfer between the initial and final states, q. The explicit result for a plane wave vector k, scattering by an angle θ appears in Eq. (6.8).

$$f(\theta) = \frac{1}{q} \int y \sin(qy) V(y) dy$$

$$= \frac{2mV_o}{\hbar^2 q^3} (\sin qa - qa \cos(qa)) \tag{6.8}$$

$$q = 2k \sin\left(\frac{\theta}{2}\right)$$

```
>> QM_Stark
  Program to look at the Stark effect in H2 for n = 1,2


n = 1 state (ground) is not-degenerate
n = 2 is degenerate, l = 0 and l = 1; 4 states
External elecric field E, along z
Perturbation ~ z is ~ Y10, no first order change in <1|z|1> ground state
Second order sum over all states ,sum<1|z|k>^2
Induced dipole moment for n = 1


        3  2
     8 ao  e
     --------
        3
n = 2 HI has m = 0, mixing of degenerate states m = 0 , l = 1,0

A =

[          0,   -3*E*ao*e]
[ -3*E*ao*e,           0]
```

Figure 6.11: Printout from the script "QM_Stark". There are two eigenvectors and two diagonal eigenfunctions for the $n = 2$ mixing of the $l = 0$ and $l = 1$ states caused by the perturbation matrix A of the interaction Hamiltonian H_I. The Bohr radius is a_o.

```
            V =

         [ -1,  1]
         [  1,  1]


            D =

         [ 3*E*ao*e,           0]
         [        0,   -3*E*ao*e]
```

Figure 6.12: Solutions of the perturbation theory eigenvalue equations for the $n = 2$ hydrogenic atom.

The three-dimensional well is of depth V_o and is located at radii less than a. In the regime where ka is much greater than one the scattering cross section has a strong forward maximum at angles less than $1/ka$ which is typical for the "diffraction" of classical waves.

The square well scattering in three dimensions is covered in the script "QM_Scatt_3dwell". Printout is shown in Figure 6.14.

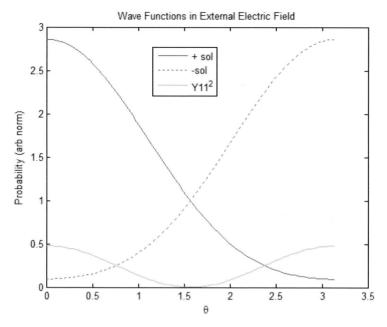

Figure 6.13: Plot of the unperturbed wavefunction $|2, 1, 1>$ and the eigenvectors of the perturbation $|2, 1, 0> + |2, 0, 0>$ where the numbers define the states n, l and m. The unperturbed state is symmetric about $90°$.

The scattering amplitude is printed, the differential cross section, and the total cross section all of which are solved for symbolically.

The differential cross section for $ka = 1$ is shown in Figure 6.15. The distribution is forward peaked but also somewhat isotropic. The cross section for $ka = 10$ is shown in Figure 6.16. In this case there is a sharp forward peak with characteristic angles about 0.1 radians or 5.7 degrees. This behavior is similar to that for classical diffraction of waves.

6.7. Photoelectric Effect — Continuum

The photoelectric effect refers to the process whereby an incident photon interacts with a bound atomic electron, is absorbed, and the final state electron is in a higher energy state. In the present case it is assumed that the final state is unbound, in the energy continuum. In this case the final state $|f>$ can be approximated as a plane wave.

```
>> QM_Scatt_3dwell
   Program to look at the scattering by a 3-d square well , r = a


Wave with vector k scatters off 3-d well, depth Vo r = a, Born approx
Scattering amplitude, q = 2*k*sin(th/2)

f =

(2*Vo^m*(sin(a*q) - a*q*cos(a*q)))/(hbar^2*q^3)

Scattering at high energies confined to theta < 1/ka
Angular distribution, dsigmadqa

                    2   2                              2
            8 pi Vo  a m  (sin(a q) - a q cos(a q))
           ---------------------------------------------
                                4    2 5
                             hbar  ka  q
Total scattering cross section

                    /                        2                              \
                    |              2   2   sin(2 a ka)      a ka sin(4 a ka)|
                    |             a ka + ------------- -   ---------------- |
            2  6  2 |   1                      4                  2         |
        8 pi Vo  a m|  ---  -  -----------------------------------------    |
                    |   4 a                         5    4                  |
                    \                           16 a  ka                    /
                   -------------------------------------------------------------
                                          4   2
                                       hbar  ka
```

Figure 6.14: Printout from the script for Born scattering off a 3d square well.

In the approximation that the initial binding energy is small with respect to the initial photon energy and the final electron energy the continuum electron takes all the initial photon energy.

$$\hbar\omega \sim \frac{p_e^2}{2m} \quad (6.9)$$

The cross section for this process, $d\sigma/d\Omega\, d\omega$, is evaluated in the script "Continuum_pe_effect". Printout from that script appears in Figure 6.17. The atom is assumed to be initially in the ground state, with a wavefunction which is hydrogenic except that the full Z attractive protons are assumed to be active. No screening of the

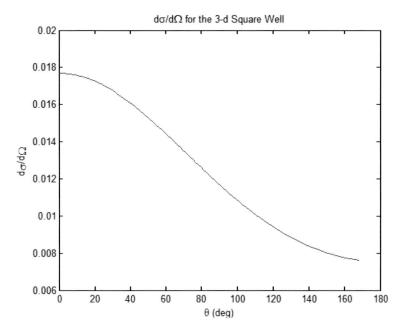

Figure 6.15: Angular distribution for scattering off a 3d square potential with $ka = 1$.

nuclear charge by the other electrons is assumed, so that the Bohr radius, a_o, is reduced by a factor Z.

The user supplies values for the parameters Z and the photon energy in $m_e c^2$ units. Results for the choices $Z = 90$ and $\hbar\omega/m_e c^2 = 0.01$ and 10, 0.4 are shown in Figures 6.18 and 6.19, respectively. In the first case the angular distribution is a dipole like form reflecting the transverse nature of the electric fields associated with the photon. In the second case the electrons are thrown forward in a relativistic effect, the "searchlight" effect.

The dependence on the incident photon energy and the atomic number Z is shown in the contour plot of Figure 6.20. The energy dependence is not very strong, but the Z dependence shows a strong enhancement at low Z. This effect arises because the "target" is smaller at high Z since the ground state radius for an inner electron is reduced by Z because of the strength of the attraction of the Z protons.

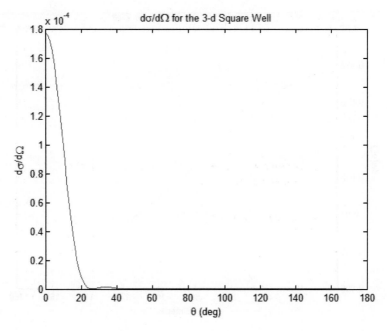

Figure 6.16: Angular distribution for scattering off a 3d square potential with $ka = 10$. In this case there is a sharp, forward, diffractive peak.

```
>> Continuum_pe_effect
   Program to explore the photo effect with e in the continuum


   Atom, with Z in ground state, r ~ ao/Z
   Time dependent perturbation theory
   e/m(A*p) is the pert, <f|x|i>
   dsigma/domega*dw with hbar*omega = hw in mec^2 units

              1/2   5            8    2   1/2           2
             32 2      Z    alpha   ao   hw       sin(t)
   --------------------------------------------------------
            2      2              1/2   1/2              4
        (- Z  alpha  + 2 hw (2      hw     cos(t) - 1))
```

Figure 6.17: Printout which shows the differential angular distribution as a function of the photon energy, where alpha is the fine structure constant, and t is the polar angle.

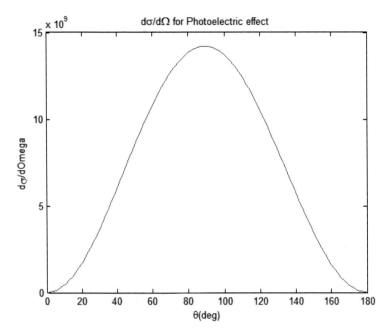

Figure 6.18: Angular distribution for the case where Z is assumed to be 90 and $\hbar\omega/m_e c^2$ is taken to be 0.01 and the process is in the non-relativistic regime.

6.8. Line Width

The wave function in quantum mechanics, ψ, can be taken to be a function of momentum or position or energy or time because they are complementary variables. One can freely convert from one description to the other, such as conversion from time, t, to energy, E.

$$\Psi(E) \sim \int_{-\infty}^{\infty} \Psi(t) e^{iEt} dt \qquad (6.10)$$

Suppose that the state being described decays in time with a lifetime τ. The reciprocal of the lifetime is called the width of the quantum state, Γ because any unstable state will have a spread in energies of a magnitude about the width. Expressed in the energy variable, the state has a spread in energies due to the uncertainty principle and since the time of the state's existence is limited.

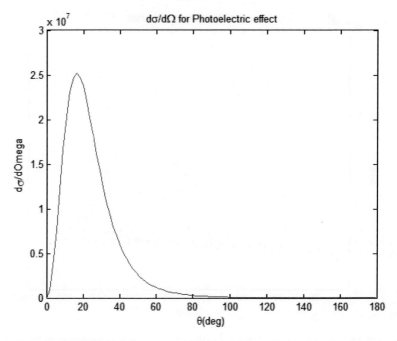

Figure 6.19: Angular distribution for the case where Z is assumed to be 10 and $\hbar\omega/m_e c^2$ is taken to be 0.4.

That spread is called the natural width, which has a Lorentzian shape when described using energy as the variable. In the time domain the wave function is an exponential function.

$$|\Psi(t)|^2 \sim e^{-t/\tau} = e^{-\Gamma t}$$
$$\Psi(E) \sim \frac{1}{\left(E + \frac{i\Gamma}{2}\right)} \quad (6.11)$$

A script for atomic electric dipole decays called "line_width" is provided. For example, the decay of the hydrogen state $2p \to 1s$ (or $n = 2, l = 1$ to $n = 1, l = 0$) has a lifetime of 1.6 nsec and emits a photon of 10.2 eV energy. The natural width is a fraction 4×10^{-8} of the photon energy, very narrow indeed scaling as approximately the cube of the fine structure constant, $\alpha = 1/137$. Printout from the script appears in Figure 6.21.

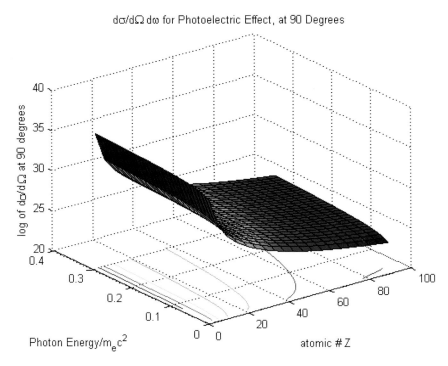

Figure 6.20: Contour plot of the continuum photoelectric angular distribution as a function of atomic number Z and incident photon energy.

The experimentally measured width as a function of temperature is shown in Figure 6.22. The natural width is temperature independent, but very narrow. The total width is dominated by the Doppler broadening of the spectrum due to thermal motion of the hydrogen gas until low temperatures are achieved in the laboratory. In fact, only fairly recently have the natural atomic widths been measured directly for a variety of gas molecules. As seen in Figure 6.22, the temperature when the natural atomic width is comparable with the thermal Doppler broadening is approximately one 100th of a degree absolute.

6.9. Barriers

The problem of a series of barriers appears in solid state physics for ionic crystals where they form a periodic potential. It can be

```
>> line_width
   Atomic Line width and Doppler broadening


Wave Function can be in x or p space or t or E variables
Decaying state, width g in t is Lorentz shape in E

psit =

exp(-g*t/2)


       /                      /  g t \               \
      | 2 exp(e t i) exp| - --- |             |
      |                      \  2  /                |              2
limit|  ------------------------------------, t = Inf | - ---------
      \           2 e i - g                   /          2 e i - g
Atomic Electric Dipole Decays, Lifetimes ~ 1 nsec
Line Width g is FWHM in E, g is hbar/lifetime
Width/Photon Energy ~ alpha^3
Pick a Lifetime (nsec), 2p to 1s in H2 1s 1.6 nsec : 1.6
Pick a Photon Energy (eV),  2p to 1s in H2 1s 10.2 eV: 10.2
Decay Width (eV) = 4.1e-07 , Width/Photon E Ratio = 4.01961e-08
```

Figure 6.21: Printout from the script "Line_width". The transformation of the wave function from the time domain to the energy domain is performed symbolically using MATLAB "int".

approximately solved. There is assumed to be a region of zero potential, where the solutions are plane waves and a region where the potential V_o is greater than the energy E so that there the solutions are exponentially damped in that region. Therefore, the electrons are bound to the crystal. The coefficients at an interface between the two regions can be found by requiring the continuity of the wave function and its first derivative. There is oscillatory behavior with wave vector k and exponential behavior with coefficient K. The incident waves with left going and right going coefficient C and D connect to $-x$ and $+x$ exponentials with coefficients A and B as;

$$\begin{bmatrix} A \\ B \end{bmatrix} = \frac{1}{2} \begin{bmatrix} \left(\frac{1+iK}{k}\right)e^{Ka+ika} & \left(\frac{1+-iK}{k}\right)e^{-Ka+ika} \\ \left(\frac{1-iK}{k}\right)e^{Ka-ika} & \left(\frac{1+iK}{k}\right)e^{-Ka-ika} \end{bmatrix} \begin{bmatrix} C \\ D \end{bmatrix}$$

(6.12)

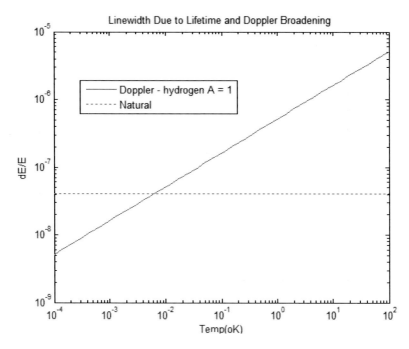

Figure 6.22: Plot of the approximate energy width for hydrogen in the $2p \to 1s$ transition. The Doppler shift due to atomic thermal velocities dominates except at low temperatures.

The script "Barriers2" allows the user to set the geometry of the array of potentials whose number is also to be chosen. The solution assumes the barriers are quite "thin", so that the exponent Ka is set to zero in the argument of the exponential. The barrier height is fixed at 7 eV.

Results for two "thin" barriers are shown in Figure 6.23 while the transmission coefficient for three barriers is displayed in Figure 6.24. In this case the resonances are quite sharp at low energies where E is less than the potential. The user is encouraged to explore a single barrier and also an array of barriers where the exponential region is less "thin". For example, a single barrier with a region of size 4 Angstroms is shown in Figure 6.25 and looks rather like the familiar transmission result for a single barrier. Looks approximately like the solution of a barrier.

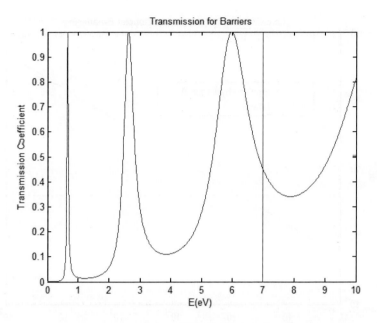

Figure 6.23: Transmission coefficient for 2 thin barriers, 1 Å wide, separated by a 6 Å free space. The barrier height is 7 eV indicated by the red line.

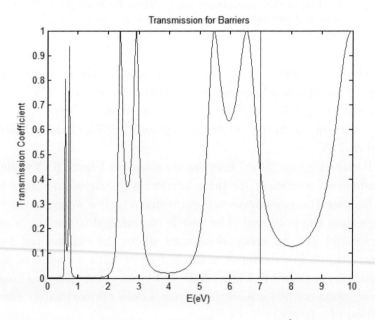

Figure 6.24: Transmission coefficient for 3 thin barriers, 1 Å wide, separated by a 6 Å free space.

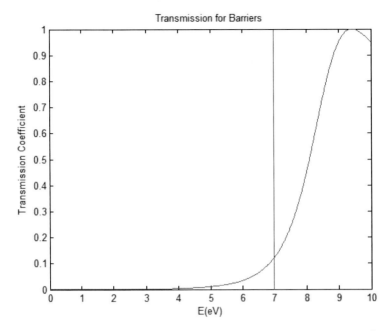

Figure 6.25: Transmission coefficient for a single barrier of width $4\,\text{Å}$ as a function of incoming wave energy E for a barrier of height $7\,\text{eV}$.

6.10. General Eigenvalues

The time independent Schrödinger equation can be solved numerically for any potential configuration. Examples are shown in the script "Gen_Eigen2" for a single well, two wells, the simple harmonic oscillator and a hyperbolic sin confining potential.

The Schrödinger equation on a numerical grid with x labeled by index j becomes:

$$\frac{d^2\psi}{dx^2} = \left[\frac{(V(x) - E)2m}{\hbar^2}\right]\psi$$

$$\psi_{j+1} - 2\psi_j + \psi_{j-1} = \left[\frac{(V_j - E)2m}{\hbar^2 \Delta x^2}\right]\psi_j$$

(6.13)

In this formula the grid spacing is Δx. The grid form of the second derivative was already seen when the discrete "slinky" was

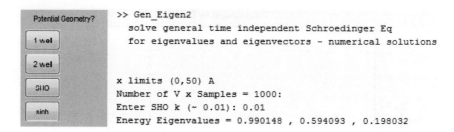

```
>> Gen_Eigen2
    solve general time independent Schroedinger Eq
    for eigenvalues and eigenvectors - numerical solutions

x limits (0,50) A
Number of V x Samples = 1000:
Enter SHO k (~ 0.01): 0.01
Energy Eigenvalues = 0.990148 , 0.594093 , 0.198032
```

Figure 6.26: Menu for the script "Gen_Eigen2". The printout for a specific SHO gives the first 3 eigenvalues.

taken to the continuum limit when the wave equation was "derived" in Section 3.

The equation could be solved using the MATLAB partial differential equation utility for a time independent special case. Alternatively, it is approached here as a grid or discrete solution. The span of x is set to be 50 Angstroms sampled 1000 times. The solutions are obtained using the MATLAB utility "eigs" with the sparse matrix, the Hamiltonian matrix which is almost diagonal. As can be seen from Eq. (6.12), the matrix has only 2×2 off diagonal terms. The time independent Schrödinger equation leads to finding the energy eigenvalues of the Hamiltonian matrix.

A menu choice is available to the user as shown in Figure 6.26 along with the printout in the specific case of a SHO. The script prints the first three eigenvalues, largest first.

The results for a specific two well potential shape is shown in Figure 6.27, while the eigenfunctions are plotted in Figure 6.28. The eigenvalues are 1.62, 0.51 and 0.46 eV in the two well example. The wells are two electron volts "deep".

Other choices are possible. In fact, it should be very simple for the interested user to edit the script and set up any problem of interest. The possibilities are open ended.

The eigenfunctions for a hyperbolic sin potential are displayed in Figure 6.29. The number of oscillations is proportional to the momentum and larger numbers lie higher in energy. That behavior in similar to that displayed by the simple harmonic oscillator.

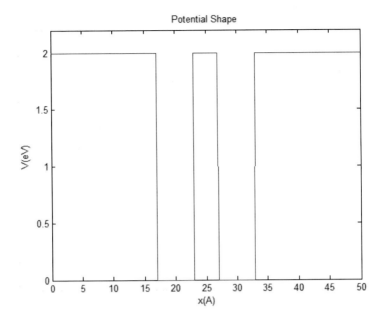

Figure 6.27: Plot of the potential in the specific two well menu choice.

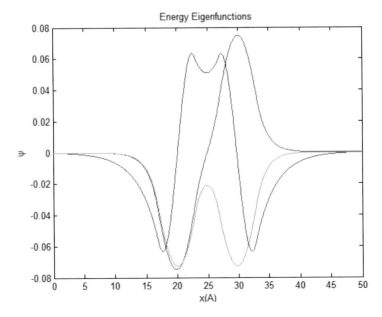

Figure 6.28: Plot of the eigenfunctions for the first three solutions in the example of the specific two well problem. The green function is the most deeply bound, followed by red and then blue.

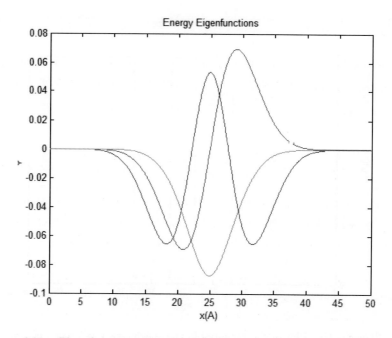

Figure 6.29: Plot of the eigenfunctions for the case of a hyperbolic sin potential. The green function is the most deeply bound, followed by red and then blue. The solutions are like those of the SHO.

Note that the "eigs" script orders the results with the highest energy first, which means that asking for the first three eigenvectors is ordered with blue as the highest energy, followed by red and then green. The user can choose parameters defining all the other menu choices.

6.11. Koenig–Penny

The Koenig–Penny model is a simplified, one-dimensional, constant and periodic potential approximation to the electron wavefunction in a solid where the ions are fixed and provide the binding forces. The wave function is assumed to be the free particle plane wave, wave vector k, modulated by a periodic function $u(k)$ which accounts for the fixed ions.

The width of the potential V_o is b, while the spacing between the regions of constant potential and of zero potential is a. Requiring continuity of the wavefunction and its first derivative at both boundaries, x of zero x of a, results in the constraint that the four homogeneous equations for u and du/dx matching have determinant zero. The solution is k, while the wave vectors for region a is k_a, while that for region b is k_b.

$$0 \sim \cos(k(a+b)) - \cos(k_a a)\cosh(k_b b)$$
$$-\frac{\sin(k_a a)\sinh(k_b b)(k_b^2 - k_a^2)}{(2k_a k_b)} \quad (6.14)$$

This constraint requires a numeric solution, and the MATLAB utility "fminsearch" is used to find k in the script "Koenig_Penny2". An approximate solution for narrow potentials is given in many textbooks, where there is a forbidden zone for the energy where no solution exits. The zone occurs approximately when $k_a a = n\pi$. The dialogue for an example with $a = 4$ and $b = 1$ Angstroms with a potential of 5 eV is shown in Figure 6.30. The energy solutions are plotted in Figure 6.31 It is notable that there are forbidden energies, near which the "fminsearch" results are a bit numerically unstable because of the discontinuities while "fminsearch" assumes the function to be minimized is well behaved. Nevertheless, the predictions of bands of energy where the wavefunctions are allowed and move freely and energy bands where they are excluded are predicted. This is basically the simplest model for the energy band theory of solids.

```
>> Koenig_Penny2
   solve 1-d periodic potential, exact
   ions ~ constant potential

Electron of Energy < Vo, Bound in a Potential Vo Outside (0,a) Where V = 0
Periodic Potential = Vo for x  =(-b,0) and x = (a,a+b) etc.
Enter Width of V = 0 (A): 4
Enter Width of Vo (A): 1
Enter Potential Vo (eV): 5
Approximate First Forbidden Energy (eV) = 2.41429
Pick an E: 1.5
```

Figure 6.30: Printout for a specific choice of the geometry and energy for the barrier height.

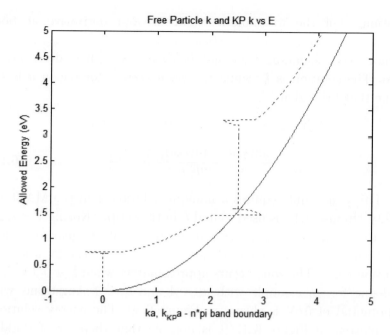

Figure 6.31: Relationship between energy, E, and wave vector, k, for a free particle (blue) and for a particle bound into a one-dimensional periodic, constant potential (red). Energies in a band approximately 1 eV wide near 2.5 eV are not possible.

The wavefunctions are required to be periodic, $u(a) = u(-b)$. The assumed form of the wavefunctions for the regions with zero potential and for binding potential, $V_o > E$, are shown in Eq. (6.15)

$$\psi = u(x)e^{ikx}$$
$$u(x) = Ae^{i(k_a-k)x} + Be^{i(k_a+k)x} \qquad (6.15)$$
$$u(x) = Ce^{(k_b-ik)x} + De^{-(k_b+ik)x}$$

Having found the wave vector k, the wave functions can be solved for at any particular, user chosen, energy. For a specific energy in the allowed band, 1.5 eV, the wavefunctions are shown in Figure 6.32. The exponential region for $x = (-b, 0)$ and the oscillatory region, $x = (0, a)$ match at $x = 0$. The regions where the potential exists

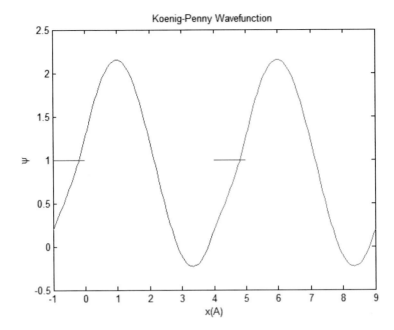

Figure 6.32: Wave functions for the specific case of parameters chosen for Figure 6.30. The function is periodic and continuous across the 2 regions of potential.

are indicated in red. The wave functions are continuous and periodic. The solutions are derived from the matching conditions, assuming A equals one, and using three of the equations to solve for B, C and D using the MATLAB matrix inversion utility, "inv", to solve the three equations in three unknowns by matrix multiplication.

6.12. Decay Chain

In the radioactive decay of an initial state, there may be several sequential "daughters", whose abundance is then time dependent. For example, $U^{238} \rightarrow Th^{234} \rightarrow Pa^{234}$. A simple case, assuming a single daughter species for each parent, appears in the script "Decay_Chain" where the species a decays into species b and hence b decays into species c. The decay rates are respectively f, g and h for species a, b and c. The first order differential equations appear in

```
Decay Chain - Coupled Rates


Decay Chain - a -> b -> c with rates f, g and h

exp(-f t)

    f (exp(-f t) - exp(-g t))
-   --------------------------
              f - g

  f g exp(-f t)         f g exp(-g t)        f g exp(-h t)
  -------------    -    -------------    +   -------------
  (f - g) (f - h)       (f - g) (g - h)      (f - h) (g - h)
Enter the a Decay Rate (~1): 1
Enter the b Decay Rate (<1): 0.6
Enter the c Decay Rate (<1): 0.5
```

Figure 6.33: Symbolic solution of the populations for the 3 species, a, b and c.

Eq. (6.6). The species a simply decays exponentially.

$$\frac{da}{dt} = -fa$$
$$\frac{db}{dt} = fa - gb \qquad (6.16)$$
$$\frac{dc}{dt} = gb - hc$$

Species b and c are "fed" by species a and b respectively while decaying with rates g and h. The equations are solved symbolically using the MATLAB utility "dsolve". The printout of that script appears in Figure 6.33. The user can change the rates. For example, it is possible to depress the population of species c by reducing the rate of species b which is the necessary "gateway" for species c. A specific example appears in Figure 6.34.

6.13. Casimir Effect

The Casimir effect is due to quantum fluctuations in the vacuum. Classically, two parallel plates with no external electric field applied

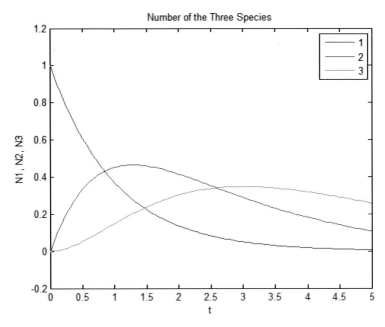

Figure 6.34: Populations of the 3 species in the specific case for decay rates of 1, 0.6 and 0.5. Species 2 builds up first, followed by species 3, while species 1 decreases exponentially.

experience no force. However, there are quantum zero point fluctuations, $\hbar\omega_n/2$ for a harmonic oscillator in the ground state, that induce a force on the plates. The fields can be imagined as arising from the vacuum fluctuating into an electron–positron pair which then goes back into the vacuum on a time scale consistent with the uncertainty principle.

This effect has been measured and has physical implications. It is, as will be seen, important only at very small plate separation. In fact even today, as microelectronics feature sizes continue to decrease, the Casimir effect must be taken into account by chip designers. This is truly an impact of quantum virtual pairs of particle-antiparticle in what is now everyday life.

One estimate of the size of the effect is to sum all zero point energies subject to boundary conditions that the wave functions vanish at the surface of the conducting plates. The quantized frequency for plates normal to the z direction, with separation d is:

$$\omega_n = c\sqrt{k_x^2 + k_y^2 + \left(\frac{n\pi}{d}\right)^2} \qquad (6.17)$$

```
>> QM_Casimir
   Program to define the size of the Casimir effect
   Energy in the vacuum in QM

   Enter plate spacing d(A): 5
   Classically no external field - no force on plates
   In QM there are vacuum fluctuations - force exists
   standing waves, kn = n *pi/d
   Zero point energy/area = hbar*integral dkx*ky*sum(wn)
   Attractive, E/A ~ 1/d^3, Pressure ~ 1/d^4
    Vacuum Energy / area (eV/A^2) = -0.219325
```

Figure 6.35: Numerical results for a user defined input of plate separation.

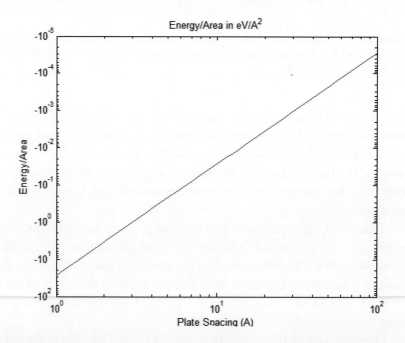

Figure 6.36: Energy per unit plate area as a function of plate spacing d, showing the inverse cubic dependence on d. With a spacing of 5 Å the energy per area is -0.22 in eV and A units.

The sum of all n and the integration over the transverse wave vectors, however, diverges and convergence factors, in this first quantization treatment, must be applied. The result is an attractive force between the plates which creates a pressure, or energy density, of:

$$\frac{\hbar c \pi^2}{240 d^4} \quad (6.18)$$

The pressure is, numerically, an energy density of $27.1\,\text{eV}/\text{A}^3$. The Casimir effect is explored in the script "QM_Casimir". Printout from that script appears in Figure 6.35.

The Casimir pressure as a function of separation is shown in Figure 6.36. It is clear that only on the Angstrom scale is the effect going to be important. However, it has been measured and it shows that there is quantum energy in the "vacuum".

Chapter 7

Astrophysics

"Astronomy compels the soul to look upwards and leads us from this world to another."
— **Plato**

"And yet it moves."
— **Galileo Galilei**

"I ask you to look both ways. For the road to a knowledge of the stars leads through the atom; and important knowledge of the atom has been reached through the stars."
— **Sir Arthur Eddington**

7.1. Transfer Orbit

The solar system offers many planets to study. As a first example a movie is provided by the script "Mars_More3" for the orbits of the inner planets Mercury, Venus, Earth and Mars. In Figure 7.1 is shown the locations of the four inner planets of our solar system at the end of a two-year time span, the last frame of the movie, when all the planets are started with an azimuthal angle of zero.

The visual experiment of watching the correlation of the orbital radii and the periods is illuminating. For a circular orbit the period τ, the energy E and the angular momentum L for an orbital radius arc, taking the planetary mass equal to one;

$$\tau = \frac{2\pi a^{3/2}}{\sqrt{GM}}$$

$$E = -\frac{GM}{2a} \qquad (7.1)$$

$$L = \sqrt{GMa}$$

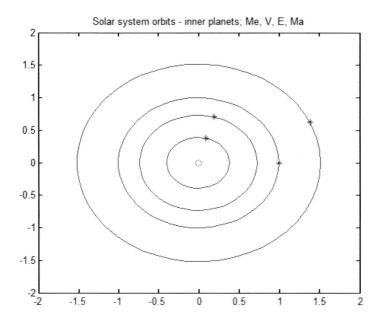

Figure 7.1: Positions of the 4 inner planets, red stars, after a time span of 2 years. Orbits are in blue and the sun is a green O.

```
The Solar System Display
Circular orbit (AU) and Period (year) for Me,V,E,Ma,JS,U,N
Orbits Connecting Earth to Outer Planets
 Period (yrs) 1.87398, Velocity (km/sec) 24.1, Radius (AU) 1.52
 Transfer Orbit; a(major) 1.26, b(minor) 1.23288, eccentricity 0.206349
 One Way Trip Time (years) 0.707173
 Period (yrs) 84.1302, Velocity (km/sec) 6.8, Radius (AU) 19.2
 Transfer Orbit; a(major) 10.1, b(minor) 4.38178, eccentricity 0.90099
 One Way Trip Time (years) 16.0492
```

Figure 7.2: Printout from the script "Mars_More3" for choices of Mars and Neptune as outer planets to travel to.

Therefore, if we specify the radius we know everything about the orbit. The dialogue printout is shown in Figure 7.2. The user chooses an outer planet.

In planning for planetary exploration it is useful to establish a feeling for the size of the times and velocities involved. For example, Mars has a period, velocity and orbital energy with respect to Earth of 1.87, 0.81 and 0.66, respectively. A "transfer orbit" starts with a

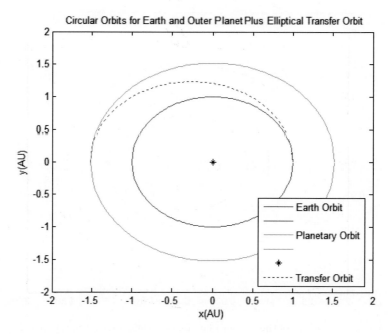

Figure 7.3: Approximate circular orbits for Earth (red) and Mars (green) and the elliptical transfer orbit between them.

velocity increase to put the vehicle into an elliptical orbit, semi-major axis $a = r_1/(1-e)$, where e is the eccentricity. The eccentricity of this orbit is $(r_1 - r_2)/(r_1 + r_2)$ where r_1 and r_2 are the turning points, the perigee and apogee.

In Figure 7.3. the radii refer to the orbital size of the earth and Mars in astronomical units, (AU). The eccentricity of the transfer orbit in this case is 0.21. The launch is at transfer orbit perigee and the rendezvous is at apogee. The velocities during the trip are shown in Figure 7.4. As can be seen, boosts in velocity are needed both at launch and at the end of the trip. The one way trip time is 0.71 years. Faster orbits are possible but more costly in terms of the fuel required and the reduced payload ratio. Clearly, there is also a "launch window" since Mars needs to be there at the end of the transfer orbit.

For other planetary trips the size and time scale gets quite large. For example, a trip to Neptune along a transfer orbit has an

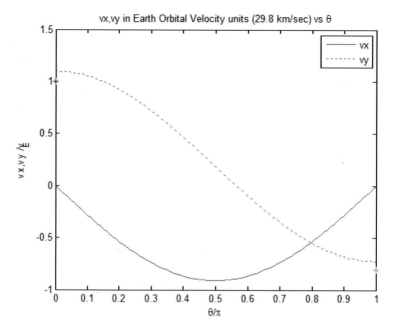

Figure 7.4: Velocities for the Mars transfer orbit. A boost is needed to begin. The red and green stars show the orbital velocities for the Earth and Mars, respectively.

eccentricity of 0.90 and takes 16.0 years. The orbits are shown in Figure 7.5. To arrive at Uranus requires a very elliptical orbit. The user is encouraged to try a few more of the outer planets which are available on the menu given by the script "Mars_More3".

It is of interest to speed up the trips. Firstly, the passengers will thank you for the speedy trip. In addition, there are fierce radiation fields to which the passengers will be exposed during the trip. Shielding would be heavy, so perhaps the orbit is not optimal for speed. Indeed, it is a minimal energy orbit, but other options cost additional fuel. At present, this problem is not fully solved.

7.2. Flyby

The velocity for an object to escape the gravity field of the earth is 11.2 km/sec, while for the sun it is 42.1 km/sec. A more economical way to escape the solar system than constructing an enormous rocket

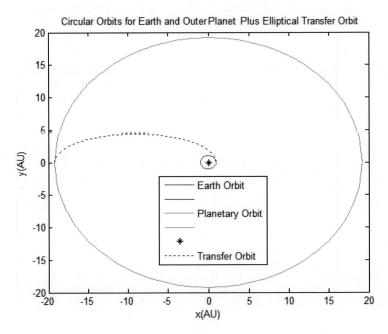

Figure 7.5: Approximate circular orbits for Earth (red) and Neptune (green) and the highly elliptical transfer orbit between them.

has been used by making a "gravity assist" or "slingshot" around the outer planets, such as Jupiter. In this fashion the Voyager probes escaped the solar system. The basic idea is that in an encounter with a very massive object, the spacecraft will essentially be dragged along and pick up a maximum of approximately twice the velocity of the object. In this case the momentum the object picks up is extracted from the gravity field of Jupiter. No actual collision occurs. The script uses a very simple numerical integration of the force on the rocket due to a planet with fixed orbit, unaffected by the spacecraft.

The printout of the script, "Jupiter_Flyby" appears in Figure 7.6.

The script supplies a movie of the encounter. The final frame of a specific example appears in Figure 7.7. On the time scales in question, the orbit of Jupiter is approximated as an unperturbed straight line with a velocity defined to be one and to be constant. The trajectory of the spacecraft is shown as a blue dashed line.

7. Astrophysics

```
>> Jupiter_Flyby
   Jupiter flyby - gravity assist


 Spacecraft Use Jupiter for a Gravity Assist to Gain Speed - Escape
   Spacecraft Picks up ~ Twice the Jupiter v, 26 km/sec
   Escape Velocity for Earth = 11.2 km/sec, for the Sum = 42.1 km/sec
   Jupiter Starts at x = -10, y = 0, x velocity = 1.0
   Spacecraft is initially at x = 0, y = -1
   Enter Spacecraft Initially vx = 0, y Velocity : 0.2
```

Figure 7.6: Printout of "Jupiter Flyby". The user defines the closeness of the encounter of the spacecraft with a point like Jupiter by choosing the initial velocity.

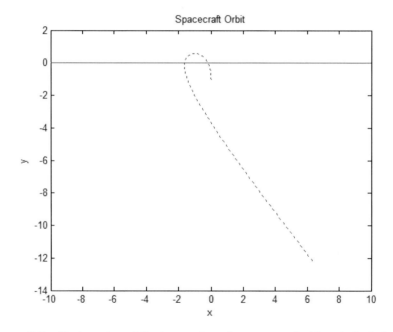

Figure 7.7: Trajectories of Jupiter, red, and a spacecraft, blue dashed, for a specific flyby which makes a close encounter between the spacecraft and Jupiter.

The success of this flyby can be seen in Figure 7.8 where the velocity of the spacecraft throughout the flyby is plotted. The y velocity, which starts at 0.2 initially, becomes -1 in the encounter which is approximately the assumed velocity of Jupiter and illustrates the "slingshot" aspect of the flyby. A simple numerical integration of the

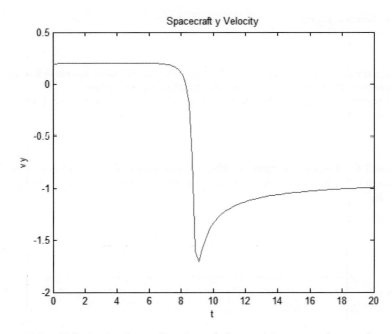

Figure 7.8: Velocity in the y direction of the model spacecraft as a function of time. The spacecraft gets a velocity increment comparable to the velocity of "Jupiter", defined to be 1, in the flyby encounter.

encounter with Jupiter's gravity is used. The actual flyby which was used trails Jupiter rather than leading it because then the direction of the spacecraft is not reversed as shown in Figure 7.9.

7.3. Lagrange Points

The use of satellites as astronomical observatories has resulted in the requirement of a stable observation platform. That is accomplished by using centrifugal force to balance gravitational attraction for a three body system. The script "Roche_Lagrange" explores the location of these stable Lagrange points for binary systems, where the user chooses the masses of the large bodies. The CM is at $x = 0$, while the masses are at positive and negative x, heavier mass to the left. The printout of the script is shown in Figure 7.10.

The total potential is the sum of the potentials of the two large bodies plus the centrifugal potential due to fact that the test body is

7. Astrophysics 203

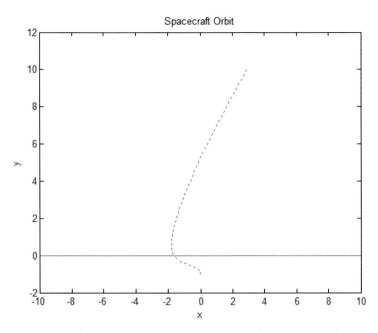

Figure 7.9: Trajectories of Jupiter, red, and a spacecraft, blue dashed, for a specific flyby which makes a close encounter between the spacecraft and Jupiter and is chosen to lag behind Jupiter, thus largely preserving the initial direction.

```
>> Roche_Lagrange
  Roche_Lagrange - 3 body problem, Earth-"Moon" and Lagrange Points


Earth-Moon System, Stable Points for a Satellite
  Examples from NASA and ESA are WMAP and PLANCK
  Balance of Gravity and Centrifugal Force
  Enter the mass ratio of Earth and "Moon" (> 1): 2
CM at x=0, m1 at -0.333333 and m2 at 0.666667
  Sort on Minimum Values for fx - numerical issues, print x ordered by |fx|
  0.666667, -1.15152, 1.23232, -1.11111, 1.27273,
  -1.19192, 0.222222, -1.07071, 1.31313, 1.19192,
```

Figure 7.10: Printout of the script "Roche_Lagrange" with a specific mass ratio chosen for the two large masses.

assumed to rotate with the binary orbital period about the two body center of mass. The mass ratio m_1/m_2 is q. There exist five stable points for this potential.

$$V = [-q/(1+q)/\sqrt{(x-x_1)^2 + y^2} - 1/(1+q)/\sqrt{(x-x_2)^2 + y^2} \\ - (x^2 + y^2)/2 \tag{7.2}$$

There are three collinear points on the x axis joining the two masses. Two are near the lighter mass; one toward the heavier mass and one away. The third point is well beyond the heavier mass, located at negative x. There are two additional points at sixty degrees to the heavier mass and at a radius which is the same as the third Lagrange point. A contour plot of the equipotentials is displayed in Figure 7.11.

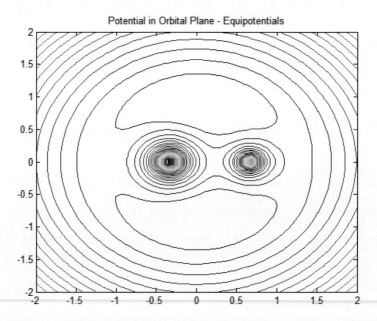

Figure 7.11: Equipotentials for the binary system with mass ratio equal to 2. The location of 3 collinear Lagrange points is visible, as are the 2 points which are at 60 degrees to the heavier mass.

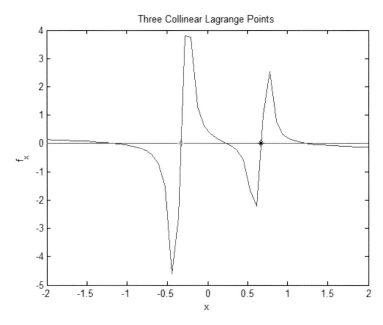

Figure 7.12: Force in the x coordinate for a binary mass ratio of 2. There are 3 collinear points. Two are near the lighter body, while the third is at large negative x, nearer the heavy body. The locations of the 2 bodies are indicated by the green and black stars.

The forces can easily be derived from the potential. The force along the x axis is plotted below in Figure 7.12. The locations of the three points where the force is zero are clear. The script prints out the locations in x for the ten minimum values of $|f_x|$. In this specific case they occur at $x = -1.15$, 1.23, and 0.22. The minima are a bit shallow, so care is needed in interpreting the use of the MATLAB utility "sort".

7.4. Binary Orbits

The simplest gravitationally bound system is a system of two objects. In the case of one very heavy object, such as the sun in the Kepler problem, the system can be taken to be a one body problem. However, in general the motion can be complex and the script to deal with it uses the MATLAB utility "ode45" to solve the problem numerically. In two dimensions the problem has eight variables; position and velocity of body1 and the same for body2. The force law is quite simple, even though the actual motion can be quite complex.

$$\vec{F}_{12} = GM_1M_2(\vec{r}_1 - \vec{r}_2)/|\vec{r}_1 - \vec{r}_2|^3 \qquad (7.3)$$

The discovery and understanding of the force law benefitted from the existence of several planets with different radii and periods and the fact that the sun has almost all the mass of the solar system. The solar system is effectively a series of one body, independent, planetary orbits as assumed previously.

One application of the general problem comes from the present search for extra solar planets. The Kepler problem solution assumes that the sun is much more massive than any planet. In that case the problem only requires the motion of the planet and the sun can be considered to be at rest. However, in general there is motion of both bodies. An example using the script "Binary2" for a separation of 1 astronomical unit (AU) with body1 having the mass of the sun and body2 having one percent of that mass is performed. The initial position and velocity of the sun are taken to be zero, while that for the "planet" has an angular velocity of a circular orbit with no radial velocity. There is a movie of the motion provided and the total paths are shown in Figure 7.13. Note that there is a wobble to the sun which could be determined using the Doppler effect to measure the velocity of the sun. This example is clearly made to enhance the effect, but nearby planets of extra-solar stars have, in fact, been observed using this technique.

At the other extreme, equal mass binary pairs can be studied. In our galaxy it appears that about ten percent of the star systems are binary, so the exercise is useful. An example with equal mass objects with a solar mass each at 1 AU is shown in Figure 7.14. For body1 the initial velocity was 6 radians/AU and the radial velocity was 2. For body2 the angular velocity was −5 and the radial velocity was 2. The resulting evolution of the binary system is complex. The reader is encouraged to play with other combinations of the six input parameters to find interesting orbits.

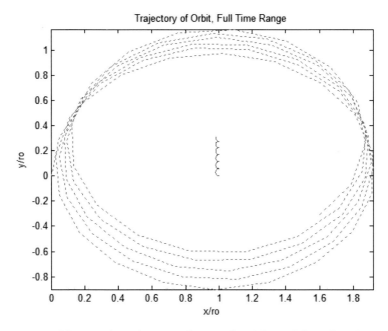

Figure 7.13: Motion of a solar mass being orbited by a "planet" with a mass of 1% of the sun. The wobble of the sun and the subsequent perturbation of the planetary orbits are very evident.

7.5. Three Body Orbits

An extension of the previous script, "Binary2", is to consider a heavy sun with solar mass that is fixed at the origin and is orbited by two "planets" in unperturbed circular orbits. The script is "Three_Body" although a general three body problem requires more variables. The forces are central, so that motion is in a plane. The user chooses a radius and mass for both "planets" with prompts for the earth and for Jupiter, as seen in Figure 7.15

The script uses the MATLAB "ode45" numerical integration od differential equations to solve for the subsequent motion of the two planets. The script is run twice, once for no perturbative force between the two planets and then again with the gravitational interaction turned on. A movie is shown in the unperturbed case. This procedure is adopted because the actual changes due to Earth-Jupiter interactions is small and even integrating over five years

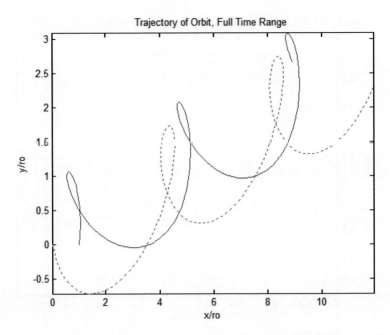

Figure 7.14: Orbital motion for equal mass binary stars. The initial conditions define roughly opposite angular velocities and equal radial velocities.

```
function Three_Body


Enter Initial Distance from the Sun for the 2 Masses(1AU, 5.2): [1 5.2]
Enter Mass of Body2, kg (me =6e24): 6e24
Enter Mass of Body3, kg (mjup = 2e27): 2e27
Velocity of circular orbit, v = 29921.6 m/sec
Velocity of circular orbit, v = 13121.5 m/sec
For circular orbit, period = 3.12883e+07 sec
For circular orbit, period = 3.71011e+08 sec
```

Figure 7.15: Dialogue for the choice of the mass and radius of 2 planets in circular orbit about the sun.

leads to small effects. The difference of the two orbits is shown in Figure 7.16, where the size of the effect is a few parts per thousand of the size of the orbit after five years. The time dependence of the difference appears in Figure 7.17.

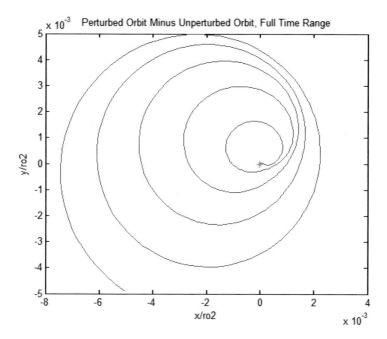

Figure 7.16: Differential effect of the perturbation due to Earth-Jupiter attraction during tracking over 5 years. Initially Earth and Jupiter are given the same azimuthal value.

A study which is less sensitive to numerical roundoff errors is to put Jupiter at a radius of two AU. In that case the perturbation of the orbit is about 2% per year relative to the unperturbed circular radius. Indeed, the interested user can explore how the radii and masses of the two planets lead to rather different orbits. For example, the heavier planet suffers smalled fractional perturbation, as expected. In extreme cases the lighted planet can be captured by the heavier planet, causing the script to terminate prematurely.

7.6. Polytropic Star

Two basic conditions for a star in equilibrium under gravitational attraction are that mass, $M(r)$ is conserved and that a pressure $P(r)$

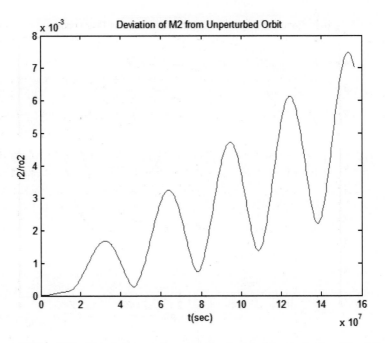

Figure 7.17: Difference between the unperturbed and perturbed orbit of Earth for 5 years scaled to the radius of the orbit of the Earth.

resists the attraction and keeps the star in equilibrium

$$\frac{dM(r)}{dr} = 4\pi\rho(r)r^2$$
$$\frac{dP(r)}{dr} = -\frac{G\rho(r)M(r)}{r^2}$$
(7.4)

The source of the pressure is not specifically specified. The Lane–Emden equation describes a form of the Poisson equation for the gravitational potential of a self-gravitating, spherically symmetric, polytropic fluid. There is assumed to be an equation of state relating pressure, P, and density, ρ, and which defines the polytrope index, n, which is a free parameter. The ideal gas law need not be assumed in this model which relates pressure and density through the index, n.

$$P \sim \rho^{(1+1/n)}$$
(7.5)

The temperature in the printout follows from the ideal gas law, $P = k\rho T/m$ where k is the Boltzmann constant and m is the mean mass of the molecules making up the star.

The Lane–Emden equation is solved numerically using the MATLAB utility "ode45". The two constraints in Eq. (7.4) are imposed with a defined relationship between pressure and density set by n, Eq. (7.5). The input needed is the core, $r = 0$, temperature. The pressure and density are derived quantities in this formulation. The solutions are integrated to larger r and terminate at R where the temperature is defined to be zero as a boundary condition. The solution is fully defined by the normalized temperature φ treated as a function of normalized radius r/R.

$$T = T(0)\phi, \quad \phi(R) = 0, \quad \phi(0) = 1$$
$$\rho \sim \rho(0)\phi^n, \quad P \sim P(0)\phi^{1+n}$$
(7.6)

Printout from the script "Star_Polytropic" is shown in Figure 7.18. The index n of 3.5 gives a reasonable representation of the solar data which is plotted on the same graph as the polytropic solution. The user chooses the index and can, therefore, see how the choice changes the density, temperature and pressure radial profiles.

The temperature profile is shown in Figure 7.19, while the pressure distribution appears in Figure 7.20. The overall agreement of this very simple model with the solar data is not too bad and is clearly a large improvement over assuming a uniform density for the star.

```
>> Star_Polytropic
   Star Model - Polytropic, index n - another model of the Sun / stars

Enter Polytropic Index: 3.5
Enter Central Density (gm/m^3), Sun is 150 - 150
Enter Core Temperature (10^7 K), Sun is 1.5 - 1.5
Core Pressure (kgm*m/sec^2) For Ideal Gas, 1.85928e+12 and Radiation 1.27575e+09
```

Figure 7.18: A printout of the dialogue for a particular, approximately, solar index.

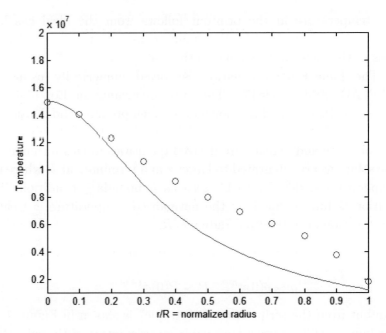

Figure 7.19: Radial distribution of temperature for a polytropic star with index 3.5. The data points, 'o', are solar data.

7.7. Pulsating Stars

It is not always the case that the hydrostatic equilibrium for a star is stable. The pressure which resists gravity is due to the elevated temperature of fusion reactions and the structure of the star may become unstable. Red giant stars are observed to pulsate, growing larger and then smaller on time scales of order months. The behavior can, in first approximation, be ascribed to a competition between the interior stellar pressure, increasing r, and the force of gravity, decreasing r. The expansion is assumed to be adiabatic, which for a perfect gas keeps $PV^{5/3}$ constant. After cooling of the shell which expands, contraction takes place and then re-heating by the core and again expansion of the shell occurs.

$$\frac{d^2r}{dt^2} = -\frac{GM}{r^2} + \frac{4\pi P_o r_o^5}{r^3} \tag{7.7}$$

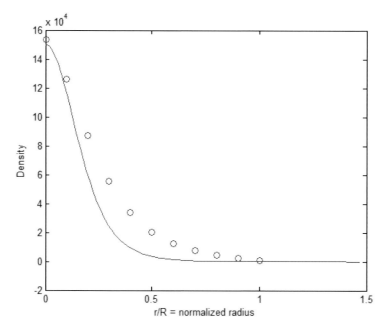

Figure 7.20: Radial distribution of the density for a polytropic star with index 3.5. The data points, 'o', are solar data.

As seen from Eq. (7.7), depending on the core pressure and radius there may be either expansion or contraction. The pulsations are considered in the script "Stellar_Pulse". The user is given the core mass, the outer shell mass and the radius. The user chooses the pressure P_0. The printout of a specific dialogue is shown in Figure 7.21.

The heat from the interior is absorbed by the shell when the hydrogen in the shell is ionized. The shell moves outward, cools, de-ionizes, and the shell is pulled back in by gravity. The balance is quite critical and values of the pressure from 0.4 to 0.6, in units of $10^5\,\mathrm{Nt/m^2}$, show stable oscillations, while outside this range the behavior is not stable.

Plots of radius as a function of time are shown in Figures 7.22 and 7.23 for pressures of 0.4 and 0.6, respectively. The pulsation period in the former case is fairly slow, on the scale of months, while in the latter case the scale is much more rapid, on the scale of

```
>> Stellar_Pulse
 Giant Stars pulsate due to inoizing and de-ionizing of outer shell

Heat from stellar interior flows outward
 Outer Shell Mass(kg) = 1e+26, Shell Radius(m) = 2e+10
 Core Mass(kg) = 1e+31
 Competition from gravity and pressure -> oscillation
 Enter the value of pressure (Nt/m^2) (10 .^5 units)): 0.6
 Radius where Pressure = Gravity, & expansion > contraction   9.61701e+10
```

Figure 7.21: Dialogue for stellar pulsations. Note that the sun has a radius of 7×10^8 m and a mass of 2×10^{30} kg. In comparison the script defines a red giant star.

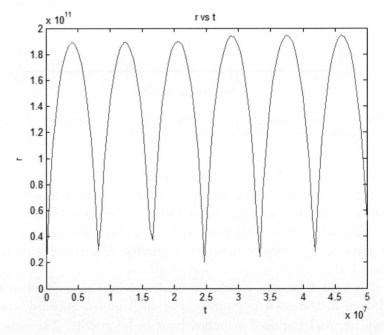

Figure 7.22: Pulsations as a function of time (sec) for a red giant with pressure 0.4×10^5 Nt/m^2.

weeks. The amplitude is much smaller in the latter case than in the former, as can be seen in the plots which are provided. Indeed, this simple picture gives an approximate model for pulsating red giant stars.

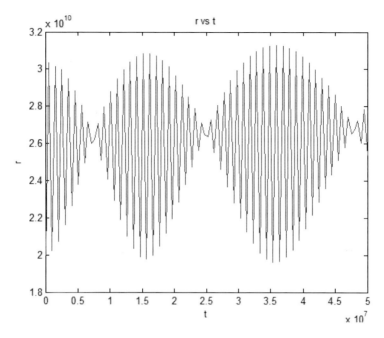

Figure 7.23: Pulsations as a function of time (sec) for a red giant with pressure $0.6 \times 10^5 \, \text{Nt/m}^2$.

7.8. White Dwarf

Stars fuse light nuclei and the resulting gas pressure and radiation pressure balance the attraction of gravity. This mechanism was implicit in the polytropic star but not explicitly modelled. However, when the final nuclei are iron, fusion is no longer exothermic and the star cools. In that case there is a contraction due to gravity which is no longer opposed by gas and light pressure but by the Fermi pressure of the electrons (fermions) in the star.

The nuclei have all the mass, but the electrons supply all the pressure. As the density increases, the star shrinks. When the electrons begin to become relativistic, the equation of state for the electrons changes. The Fermi momentum, p_f, is related to the electron number density, n. The fermions cannot occupy the same state and the number of possible states in three dimensions goes as the third power of the size of the star, V. Therefore, as the density n increases, the momentum of the fermions must increase.

The electron equation of state for pressure P depends on density, ρ, with the fermion energy going as the square of pf at low density and becoming linear in p_f at high density. The electron to nucleon fraction is Y_e, is assumed to be for iron in this script. Since pressure is an energy density, $dP/d\rho$ is dimensionless, c equals one unit.

$$p_f = (3\pi^2 n)^{1/3}, \quad n = \frac{N}{V}$$

$$\frac{dP}{d\rho} = Y_e \gamma \left(\frac{m_e}{m_p}\right), \quad \gamma = \frac{x^2}{3\sqrt{1+x^2}}, \quad x = \left(\frac{n}{n_o}\right)^{1/3}$$

(7.8)

The typical values of density are n_o, the number density in the transition region where the electron momentum becomes comparable to the mass. The total mass density at that point, ρ_o, is determined by the nuclei and leads to a typical radius R_o and mass M_o.

$$\frac{p_f}{m_e} = \left(\frac{n}{n_o}\right)^{1/3}, \quad n_o = \frac{m_e^3}{3\pi^2}, \quad \rho_o = \frac{m_p n_o}{Y_e}$$

(7.9)

$$R_o = \left[\frac{Y_e \left(\frac{m_e}{m_p}\right)}{4\pi G \rho_o}\right]$$

$$M_o = \frac{4\pi R_o^3 \rho_o}{3}$$

Numerical values of the scale factors for radius and mass appear in the printout of the script, shown in Figure 7.24.

```
>> White_Dwarf
   Model of White Dwarf - Pressure from e, Mass from Fe nuclei


   e  number density in 1/cm^3 when pf = me = 5.9e+29
   p  mass density in gm/cm^3 when e = no = 2.11207e+06
   Typical Radius in cm = 3.5728e+08 and mass in gm = 1.22073e+33
   Enter Central Density in rho 9.8 x 10^5/Ye (gm/cm^3) units : 10
   Total Star Mass (gm) = 1.55495e+33, and Radius (cm) = 4.91043e+08
   Solar Mass (gm) = 2e+33, and Radius (cm) = 7e+10
```

Figure 7.24: Printout from the dialogue of the script "White_Dwarf". Numerical values for the scale of electron number density, mass density, radius, stellar mass and stellar radius are given.

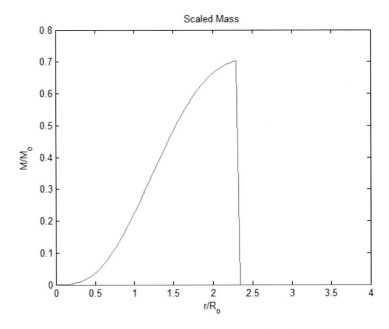

Figure 7.25: Scaled mass as a function for scaled radius for a white dwarf with a scaled central density of one.

The white dwarf is explored in the script "White_Dwarf" where the differential equations are solved numerically using the MATLAB "ode45" utility. A central density is chosen by the user and then the density is integrated in radius until it reaches zero at R, where the total mass is defined to be contained within radii less than R. A plot of the mass, scaled to M_o as a function of radius, scaled to R_o for a central density of ρ_o equal to one appears in Figure 7.25.

The white dwarf is not like a typical star, for example, the polytropic star displayed above, because the attraction of gravity is opposed only by the Fermi pressure of the electrons. A plot of the mass and radius for different central densities appears in Figure 7.26. At low densities the mass grows with density and the star contracts. At higher densities the mass levels off because the electrons are becoming relativistic and cannot support the gravitational attraction. This level is roughly at the mass of the sun. However, the radius is approximately 1/10000 the solar radius, indicating

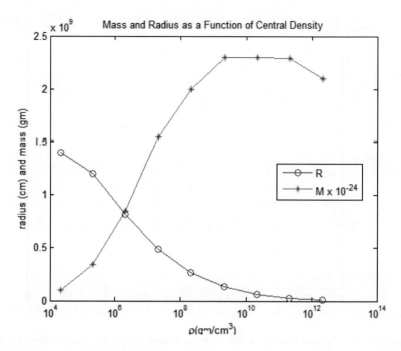

Figure 7.26: Radius and mass of a white dwarf as a function of the central, core, density of the star. The radius shrinks with density and at high density the mass levels off at values approximately one solar mass.

the compact nature of a white dwarf. Above about one solar mass the Fermi pressure of the electrons cannot resist the formation of a black hole and the heavier neutrons must take over with their pressure.

7.9. Boltzmann and the Sun

It is perhaps amusing to connect the fact that the sun is yellow with the solar constant on earth which is fundamental to solar power feasibility studies. The energy per unit area radiated by a blackbody is proportional to the fourth power of the temperature. The constant of proportionality is the Stefan–Boltzmann constant σ. The value of σ is 5.67×10^{-8} watts per square meter per degree[4]. The energy density is u. It has a distribution of energy which is $du(T)/dE = (kT)^3/\pi^2(\hbar c)^3[x^3/(e^x - 1)]$, where $x = E/kT$

$$\sigma T^4 = -\frac{\pi^2(kT)^4}{60\hbar(\hbar c)^2}$$

$$u = \frac{4(\sigma T^4)}{c}$$

$$u = \frac{(kT)^4 \pi^2}{15(\hbar c)^3} \quad (7.10)$$

The distribution of energies of the emitted photons follows the Bose–Einstein distribution, but with a zero chemical potential. The chemical potential is zero because all the photons can sit in the lowest

```
>> Boltz_Sun2
   Program to connect sun color and solar constant on Earth


Stefan-Boltzmann Law, du(T)/dE=(kT/hbar*c)^3/pi^2[x^3/(exp(x)-1)], x = E/kT
Symbolic Integration for u

     2   4
   pi  kT
   ---------
        3
   15 hbarc
Sun is "yellow" - peak emission at ~ 5700 Angstrom

             2
         16 pi  hbar
   ---------------------------
    5 /    / 2 pi hbar \     \
  lam |  exp| --------- | - 1 |
      \    \   T lam   /     /|
BB Emission spectrum, energy density per unit lambda
hbar in c = 1 units, T in k = 1 units
Enter Solar Temperature in K units ~ 5700: 5700
Total Power in Watts/m^2 ~ T^4 - Stefan - Boltzmann Law
Solar emission (MW/m^2) at solar surface = 59.8525
Solar constant (kW/m^2) on earth = 1.26648
Enter Solar Temperature in K units ~ 5700: 4000
Total Power in Watts/m^2 ~ T^4 - Stefan - Boltzmann Law
Solar emission (MW/m^2) at solar surface = 14.5152
Solar constant (kW/m^2) on earth = 0.307142
```

Figure 7.27: Printout of the script "Boltz_Sun2" for two samples of solar surface temperature. The quoted formula is for energy density per wavelength.

quantum state since the number density of photons is not constant because the system is in a "bath" at constant temperature. The density of states, or phase space for photons goes as $dP_x\, dP_y\, dP_z/\hbar^3 = 4\pi P^2\, dP/\hbar^3$ so that the spectral energy density $du(E)/dE$ goes as the cube of the energy times the Bose–Einstein statistical factor.

The emission spectrum of the sun peaks at a wavelength, yellow, that corresponds to a temperature on the surface of the sun of about 5700 K. That means a radiation of about 60 million watts per square meter. Scaling by the inverse radius of the sun and the earth radius of one AU, the solar constant for energy falling on the earth is $1.27\,\text{kW/m}^2$. The script "Boltz_Sun2" makes these calculations and invites the user to try different temperature, perhaps appropriate to red giant stars or blue stars. Printout for two examples is given in Figure 7.27. It is clear that the radiated power is very strongly dependent on the temperature.

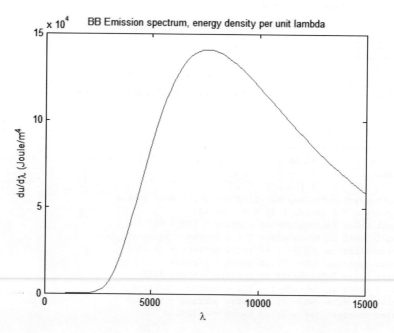

Figure 7.28: Spectral energy density for a temperature of 4000 kelvins. Wavelength is in Å units.

7. Astrophysics

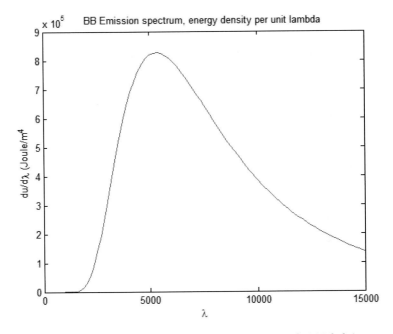

Figure 7.29: Spectral energy density for a temperature of 5700 kelvins, appropriate to the sun.

The energy density per unit of wavelength follows from a change of variables from energy to wavelength for photons and is:

$$\frac{du(T)}{d\lambda} = \frac{16\pi^2 \hbar c}{\lambda^5 (e^{2\pi \hbar c/\lambda kT} - 1)} \quad (7.11)$$

A plot of the spectral distribution is shown in Figure 7.28 for a temperature of 4000 kelvins and in Figure 7.29 for the standard solar surface temperature of 5700 kelvins.

Chapter 8

General Relativity

"God abhors a naked singularity."

— **Stephen Hawking**

"It is impossible to imagine a four-dimensional space. I personally find it hard enough to visualize a three-dimensional space!"

— **Stephen Hawking**

8.1. Light Deflection

Light is deflected by nearby masses because light has energy and energy has mass and all mass gravitates. The geodesic equation for light with the variable change $u = 1/r$ is given below in Eq. (8.1). This equation does not have a closed form solution. Nevertheless, an approximate solution for small light deflections consists of assuming a Newtonian straight line path for the term proportional to GM.

$$\frac{d^2 u}{d^2 \phi} + u = 3GMu^2$$

$$u \sim \frac{\sin \phi}{R}$$

$$\frac{d^2 u}{d^2 \phi} + u \sim 3GM \sin^2 \frac{\phi}{R^2} \qquad (8.1)$$

The approximate equation has solutions:

$$u \sim \frac{3GM \left(1 + \frac{\cos(2\phi)}{3}\right)}{2R^2} + \sin \frac{\phi}{R} \qquad (8.2)$$

The approximate solutions are displayed by the script "Light_Deflect" which uses the MATLAB utility "dsolve" as shown in Figure 8.1. There is a symmetry in the solutions before and after

8. General Relativity

```
>> Light_Deflect
   Program to look at GR light deflection


Enter rs/R: 0.01
Newtonian straight line u = 1/r = sin(phi)/R
GR not in closed form, start with u = 1/r = sin(phi)/R
Newtonian asymptotes at phi = 0, no deflection
GR asymptotes, small angle approx, at +- rs/R, total deflect 2rs/R

                                             / cos(2 t)  \              / cos(2 t)  \
                                           m |  --------  - 5/2 |    m cos(2 t) |  --------  - 5/2 |
  3 m            m cos(2 t)   m cos(4 t)     \    2       /                \    2       /
  --- + sin(t) - ---------- + ---------- - 2 m cos(t) -  ---------------------  -  ---------------------
   8                2             8                              2                           2
GR deflection, R = 1, m = GM is

asym =

4*m + 4967757600021511/40564819207303340847894502572032
```

Figure 8.1: Printout from the script "Light_Deflect". Showing the symbolic solution of the approximate equation for the trajectory of light.

the point of closest approach. In the printed evaluations, the initial conditions are a radius at infinity, u of zero at angle of zero, and an initial angular velocity of $1/R$. The light deflection angle is evaluated for a given value of the impact parameter, R with respect to the Schwarzschild radius $r_s = 2GM/c^2$.

The asymptotes can be read off from the approximate equation above for zero angle, and u of zero, $\pm 2GM/Rc^2$. The small angle solution for the angle between the asymptotes is $2r_s/R$. This appears in the printout as "4m" plus a small correction for the approximate initial conditions used, since $m = GM$ and $R = 1$ in the script. For the sun the deflection is rather small, but light passing near a black hole can have major angular deviations from a straight line. The plot made by the script is shown in Figure 8.2.

The light orbits for Newtonian and general relativistic orbits are shown in Figure 8.3. The blue line is the Newtonian straight line, while the red dashed line is for $r_s/R = 0.02$ and the black line is a distorted representation of the surface of the sun, radius R.

8.2. Circular Geodesic

In classical mechanics, any radius is possible for a circular orbit. The classical radius, a, depends quadratically on the angular momentum,

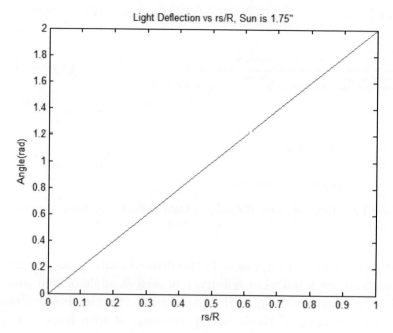

Figure 8.2: Light deflection angle as a function of the ratio of impact parameter to r_s. The result is strictly only valid for small deflections.

L, of the orbit, as mentioned in Eq. (7.1) above

$$a = \frac{J^2}{GM}$$
$$J = \frac{L}{m}, \quad L = mva \tag{8.3}$$

The orbits in a Schwarzschild space are examined in the script "circle_geodesic2". The equation for a circular orbit has a classical term plus an additional term due to general relativity for the metric, as quoted previously in Eq. (8.1), $3\,GM/u^2$. The classical equation for $u = 1/r$ has a constant of the motion L. The general relativistic geodesic has a related constant, h, which is approximately L/c. The radius of light, a_γ, can be read off Eq. (8.1), $u = 3\,GMu^2$, or

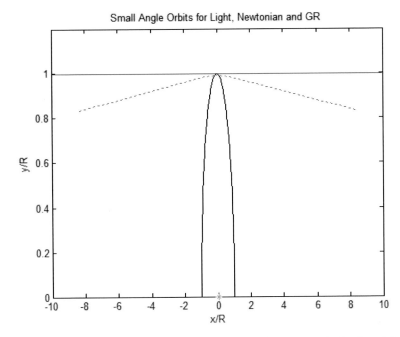

Figure 8.3: Light trajectories for Newtonian light (undeflected, blue) and light with a small ratio of the impact parameter and the Schwarzschild radius, 0.02. The black line is the solar sphere.

$a_\gamma = 3\,GM/c^2 = 3r_s/2$

$$u = \frac{r_s}{2h^2} + \frac{3u^2 r_s}{2}, \quad u = \frac{1}{r}$$
$$a_\gamma = \frac{3r_s}{2}$$
(8.4)

Light is "attracted" to the sun, and it has an infinite h which means the circular orbit for light has a radius 3/2 the Schwarzschild radius, $r_s = 2GM/c^2$. In Eq. (8.4) the first term is the classical result $u = 1/a \sim GM/(hc)^2$.

For a material particle, h is finite and the solution to the geodesic equations is a function of the parameter h which is a constant of the motion, has the dimensions of length and is approximately L/c. In the classical limit, $a \gg r_s$ and $h \gg r_s$, h approaches L/c and a

approaches L^2/GM

$$a = \frac{(h)^2}{r_s}\left[1 + \sqrt{1 - 3\left(\frac{r_s}{h}\right)^2}\right]$$
$$h = \frac{L}{c}\left[\frac{1}{\left(1 - \frac{r_s}{a}\right)}\right] \qquad (8.5)$$

In the script, the user picks an initial value of the constant of motion h. It is clear from Eq. (8.5) that the value of h must be greater than $\sqrt{3}r_s$ and therefore the radius must be greater than three times the Schwarzschild radius or twice the photon orbital radius. An example of the printout for the script appears in Figure 8.4.

The script supplies the radius as a function of the constant h. The radius, both classical and relativistic, appears in Figure 8.5. The relativistic radius approaches three times the Schwarzchild radius, while the classical radius is smaller and has no minimum. Note the photon point which has a radius twice as small as the smallest possible orbit for a material particle. However, that orbit is unstable as are all orbits for material particles below three Schwarzschild radii unless external power is applied.

8.3. General Geodesic

The Schwarzschild metric is the solution in General Relativity (GR) for the space-time due to a point mass at the origin. Particles and light travel on geodesics in such a metric. Radial geodesics have

```
>> circle_geodesic2
   Circular Geodesic - in a Schwarz space


   Schwarzschild Radius, rs = 2964.44 m
   Enter the GR Constant of Motion h (rs) > sqrt(3): 1.8
   GR Radius (rs) = 4.12182
      Classical Angular Momentum, L = 1.21243e+12 m^2/sec
      Classical Radius (rs) = 3.71717
         Classical Escape Velocity, v = 1.55602e+08 m/sec
         Classical Circular Velocity, vc = 1.10027e+08 m/sec
```

Figure 8.4: Dialogue for the circular geodesics.

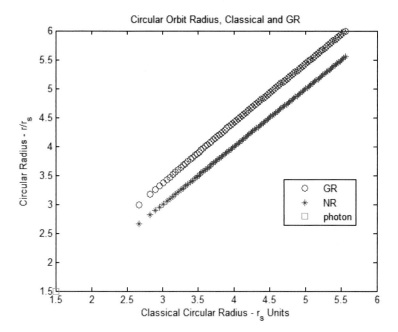

Figure 8.5: Radius of a circular orbit for a classical and a relativistic treatment. Note the GR minimum radius is 3 in r_s units. There is no classical minimum. The photon orbit at 3/2 units is also shown.

solutions which exist in closed form;

$$(r - r_s) = (r_o - r_s)e^{-ct/r_s}$$

$$s_o - s = \frac{2(r^{3/2} - r_o^{3/2})}{3\sqrt{r_s}}$$

(8.6)

The coordinate radius is r and the coordinate time, measured at large r, is t, while the proper time is s. The initial radius is r_o, the initial velocity is zero and r_s is the Schwarzschild radius. The proper time is well behaved at all r, while the coordinate time t diverges as the radius r approaches r_s. Clearly as r approaches r_s, ct must approach infinity and an object falling into a black hole takes forever to reach the Schwarzschild radius.

In the general case, analogous to the Kepler problem in classical mechanics there is a constant of motion, h, like angular momentum.

There is another choice, basically that proper and coordinate clocks at large r run at the same rate.

$$r^2\left(\frac{d\phi}{ds}\right) = h$$
$$\left(\frac{1-r_s}{r}\right)\left(\frac{dt}{ds}\right) = 1 \tag{8.7}$$

Using the constants of motion, the motion is in a plane as it is classically and this fact is used to reduce the problem to one with only one dimension. The equation of motion for $u = 1/r$ is:

$$\frac{d^2u}{d\phi^2} + u = \frac{r_s}{2h^2} + \frac{3r_s u^2}{2}$$

$$h = \frac{\left(\frac{L}{c}\right)}{\left[1 - \frac{r_s}{r}\right]} \sim \frac{L}{c} \tag{8.8}$$

$$\frac{L}{c} = \frac{r^2\,d\phi}{dt}$$

The first three terms arise in classical mechanics and define the Kepler equation. The last term is added by considering the general relativistic geodesic motion in the Schwarzchild space corresponding to a point particle of mass M. For photons, the constant h approaches infinity.

The solutions are calculated in the script "gen_geodesic" which uses the MATLAB utility "ode45". The user chooses classical, relativistic or photon solutions. The trajectory is defined by user inputs for the initial position in r, the impact parameter (phi angle) and the L/c value in m units. The orbit is plotted for an angular range also defined by the user.

The orbits for photons are almost straight lines for large r as is expected classically because the metric becomes flat at large distances. However, for smaller initial r, the deflection of light is very evident. A sample plot for light is shown in Figure 8.6.

The orbits of bound systems are not re-entrant in general relativity as are Kepler orbits. This causes the famous perihelion advance of Mercury. Setting the phi range of six (pi), the initial

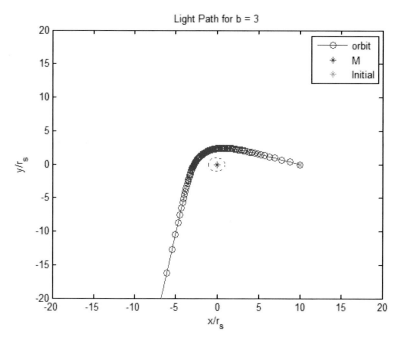

Figure 8.6: Trajectory for light with initial radius of 10 with impact parameter of 3. Note that $b = 2$ is not a stable orbit and the light falls into the origin (red star). The Schwarzchild radius is plotted with dashed red lines.

position of 5 and impact parameter of 5 with L/c of 6000 m results in the path shown in Figure 8.7. The path is a bound state for a material particle, but it is far from the familiar Keplerian re-entrant ellipse.

An unbound trajectory for initial r of ten, impact parameter of five and L/c of twenty thousand is shown in Figure 8.8. The orbit is unbound and corresponds classically to a hyperbolic orbit. The time samples are uniform so that the slowing down of the particle at large distances is observable. The relationship between coordinate time, t, and proper time on the orbit appears in Figure 8.9. At large r values the two are approximately the same, while for small radii, the proper time remains well behaved and the coordinate time is larger. This effect has, indeed, been measured and confirms the relativistic calculation.

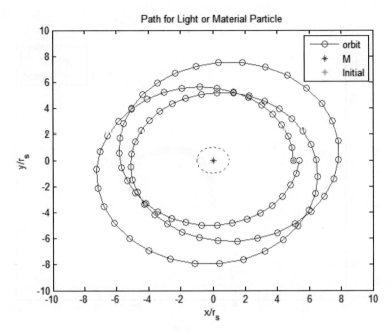

Figure 8.7: Multiple orbits for a particle started on a circular orbit at radius 5. The non re-entrant nature of the orbit is very evident in this extreme case.

Just as with Kepler orbits, the relativistic orbits have many parameters to vary. The interested user is encouraged to try many combinations and explore the subsequent motion.

8.4. Kerr Photons

The Einstein equations are non-linear because the field itself gravitates because it has energy. For this reason very few solutions to the equations have been found analytically. The first was the Schwarzschild metric and the second is the Kerr metric.

The Kerr metric is the solution for a point mass M rotating with a signed parameter a, which is proportional to the angular momentum of the mass with the sign of a defining the sense of the rotation. The L/M value is approximately a. There are two possible photon equatorial circular orbits, Eq. (8.9), with radii depending on whether the orbit is in the sense of rotation or against it. The differences

8. General Relativity

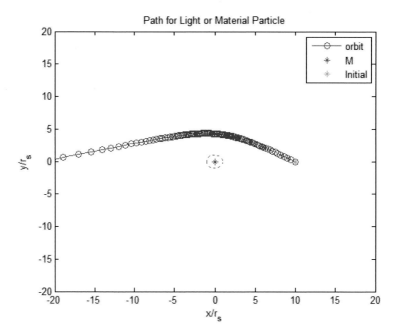

Figure 8.8: Unbound trajectory for relativistic particle motion, classically a hyperbola. At large r the particle velocity is reduced.

are due to "frame dragging" where the Kerr rotation drags the coordinates along. Spacetime is swept along by the rotation.

$$r_1 = 2M\left[1 + \cos\left(\frac{2}{3}\cos^{-1}\left(-\frac{|a|}{M}\right)\right)\right]$$
$$r_2 = 2M\left[1 + \cos\left(\frac{2}{3}\cos^{-1}\left(+\frac{|a|}{M}\right)\right)\right]$$
(8.9)

Because of frame dragging there are no purely radial geodesics. In general radial geodesics are impossible while circular geodesics are possible. There are two event horizons. The mass is M, or $GM/c^2 = r_s/2$.

$$r_\pm = M\left[1 \pm \sqrt{1 - \left(\frac{a}{M}\right)^2}\right]$$
(8.10)

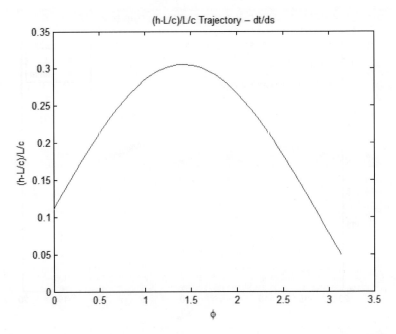

Figure 8.9: Relationship of coordinate time to proper time during the quasi hyperbolic orbit shown in Figure 8.8. At small r, there is a major difference in t and s.

The geodesic equations for photons in equatorial orbits can be cast into a simple first order differential equation. Note that a can be positive or negative, depending on the sense of the rotation. The angular variable in the equatorial plane is φ which depends on the Kerr rotation parameter a and M.

$$\frac{d\phi}{dr} = \frac{a}{(r^2 - 2Mr + a^2)} \qquad (8.11)$$

Printout from the script "Kerr_Photon2" is shown in Figure 8.10. Units with $M = 1$ are used in this script. Values for the two event horizons and the two possible circular orbits are part of the printout.

The equation for the orbit is solvable and is found using the MATLAB utility "dsolve". The symbolic solutions are also printed out, for both positive and negative a. The constants C2 and C4 depend on the initial conditions. Note that these particular photon solutions are for orbits which approach the outer event horizon for the

```
>> Kerr_Photon2
   Kerr metric - photon geodesic
   specialize to equitorial - remains planer, unique


Kerr Metric has 2 parameters - Mass and Spin
Enter a/M, M = 1, rs = 2: 0.98
Ergosphere Event Horizons, M = 1, 1.199 and 0.801003
Circular Photon Orbits, pro and retro, 1.23936 and 3.98219

phin =

C2 - (a*atan((r - 1)/(a^2 - 1)^(1/2)))/(a^2 - 1)^(1/2)


phout =

C4 - (a*atan(1/(a^2 - 1)^(1/2) - r/(a^2 - 1)^(1/2)))/(a^2 - 1)^(1/2)
```

Figure 8.10: Printout from the script "Kerr_Photon2". The equatorial photon orbits are solved for symbolically.

specific user input. Note that the trajectories are incoming toward smaller r.

An orbit for $a = 0.98$ and a minimum r of 1.25 is shown in Figure 8.11. With these parameters the outer redshift boundary occurs at a radius of 1.20 in M units. Note that the Schwarzschild radius, r_s, in these M equals one units has a value two, so that an observer can get closer to a Kerr mass than a non-rotating mass using the frame dragging spin of the Kerr metric. The frame dragging of the photon is very evident in the figure. Recall that a photon minimum radius for a circular orbit is 1.5 in r_s units, Eq. (8.4).

8.5. Kerr General

General geodesic motion in a Kerr metric is quite complicated. The script "Kerr4" is restricted to the equatorial plane because only equatorial motion remains planar due to frame dragging. Because the space is time symmetric and axially symmetric in this case, there are two constants of the motion, which can be identified with E, the energy at large radius and L, the angular momentum as in the classical Kepler problem. Units with $M = 1$ are again used,

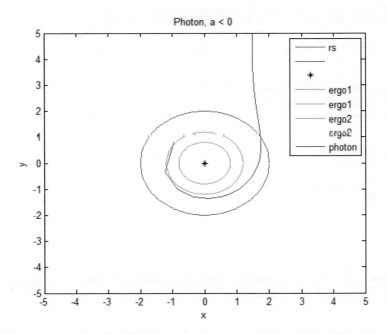

Figure 8.11: Geodesic for an equatorial photon traveling close to the outer event horizon. Units are for $M = 1, r_s = 2$.

$GM/c^2 = 1$, $r_s = 2$. There is a distance where rotation classically has a velocity of the speed of light which is L/c. As in the Kepler case, negative E orbits are bound, while positive energy orbits are usually free but may be captured by the dragging effect.

There are then differential equations for the proper time, azimuth and radius as a function of coordinate time t, with r limited to be larger than the larger event horizon. There is, in addition, a sign for the radial solutions as incoming or outgoing. The equations themselves are not very transparent, but the user can read them off from the script "Kerr4" as desired.

In "Kerr4" the user chooses the orbital parameters analogous to the Kepler parameters, E and L along with a, the initial radius, and the incoming or outgoing sense of the trajectory. An example for $a = 0.98$, $E = 1$, $L = 3$, $r_o = 1.5$ and outgoing appears in Figure 8.12. An orbit with the same parameters except the sense of a appears in Figure 8.13.

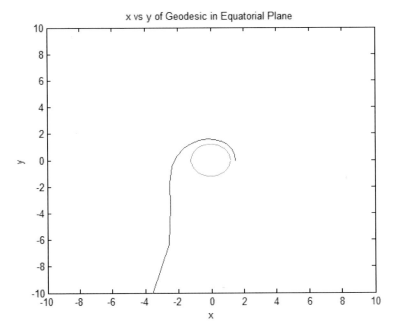

Figure 8.12: Kerr trajectory for an outgoing path, $E = 1$, $L = 3$ and initial $r = 1.5$. The green circle is the outer event horizon. The sense of the spin a and the trajectory reinforce.

Recall that Kepler trajectories are circles and ellipses for bound states and parabolas and hyperbolae for unbound states and that the plots in Figures 8.12 and 8.13 are far from the simple conic sections. Thus it is clear that the Kerr solution can have strong relativistic aspects, especially the "frame dragging" due to the spin of the mass. The solutions in Kerr4 are for material particles where L and thus h are finite. A movie is provided for the (x, y) motion along the chosen trajectory as a function of equal steps in coordinate time t.

There are also interesting bound states with negative energy. An example with $E = -1, L = 10$ initial $r = 3$ and incoming sense is shown in Figure 8.14, where the frames of the movie are plotted as open circles. In this case the frame dragging aspect of the orbit is very clear, since the sense of the trajectory changes during the orbit. The user is encouraged to play with the many available parameters

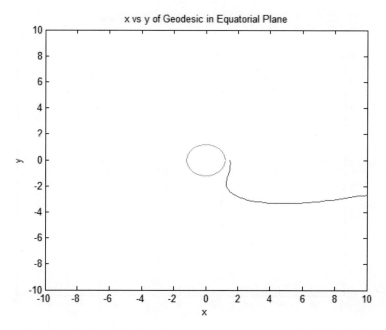

Figure 8.13: Trajectory for the same parameters as the previous figure except that the spin direction a has been reversed. The trajectory is bent toward the sense of a, reversing the initial motion.

which define a general equatorial orbit in the Kerr geometry as only a few are illustrated here.

8.6. Gravity Wave Radiation

A rotating binary star system is a good candidate to be an observable source of gravitational radiation. There are also many binary stars, so that their radiation could serve as a "standard candle" for calibrating a successful gravitational wave detector. For a binary with an individual stellar mass M and radius R, the power radiated is;

$$r_s = \frac{2GM}{c^2}$$

$$\frac{dE}{dt} = \frac{2}{5}\left(\frac{c^5}{G}\right)\left(\frac{r_s}{R}\right)$$

(8.12)

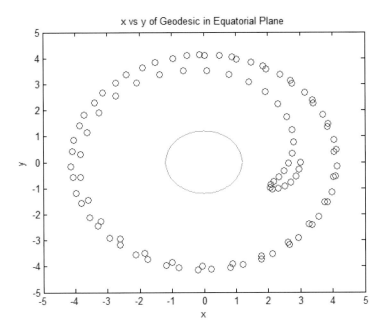

Figure 8.14: Bound state trajectory with $E = -1, L = 10, a = 0.98$ and an incoming orbit. The particle is captured by the rotation and cast into a quasi-re-entrant stable orbit.

The maximum radiated power is set by the quantity $c^5 G$ which is 3.8×10^{52} watts. The emitted gravitational waves cause a metrical distortion (dimensionless) at an observation distance r which is approximately:

$$h_{ij} \sim \frac{r_s^2}{2Rr} \tag{8.13}$$

That distortion is detected in a gravitational wave detector which can be thought of as responding to the tidal forces, Eq. (2.10), carried by the wave which elongate and compress a test body.

The distortion has a frequency of the orbital motion of the binary, which is typically in the kHz range. The quantities needed to detect gravitational waves are shown in the script "GR_Rad". If these waves can be detected, then observational cosmology can go to times shorter than the cosmic ray background decoupling time which now blocks observation since the plasma is opaque to visible light.

```
>> GR_Rad
  Program to look at GR radiation and the metrical distortion


Maximum power in gravitational waves ~ 2*c^5/5G = 1.44 x 10^52 Watts
Enter Binary in Solar mass units: 1
Enter Binary Radius in km: 100
Enter source distance in ly: 100
  Binary frequency w in Hz = 183.712
  Energy loss into radiation (W) = 4.32e+50
  Metrical distortion h = 4.75647e-17
```

Figure 8.15: Printout for a particular user choice of parameters for a rotating binary star system.

Numerical values for a user chosen M, R and r choice are shown in Figure 8.15.

Clearly, the metrical distortion is very small and sets the scale for fractional elongations in any gravitational wave detector. A contour for the orbital frequency as a function of M and R appears in Figure 8.16. The detectors must be sensitive to a range from 10^2 Hz to perhaps 10^5 Hz if they are to use binaries to calibrate. The radiated power as a function of M and R is shown in Figure 8.17 scaled to the maximum possible power. Detectors should be sensitive to metrical changes several orders of magnitude below the maximum value.

8.7. Stellar Pressure

The interior of a uniform density sphere can be explored in Newtonian and general relativistic mechanics. The Newtonian interior potential is:

$$\Phi = \frac{GM(r^2 - 3R^2)}{2R^3} \tag{8.14}$$

where M is the total mass and R is the radius. A plot of the interior and exterior potentials is shown in Figure 8.18.

The Newtonian pressure, P, can be written in terms of the Schwarzschild radius for a uniform density star. The equation of hydrostatic equilibrium was already shown for polytropic stars and

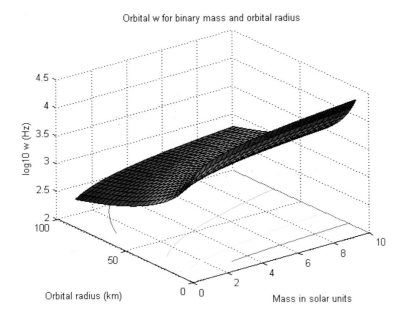

Figure 8.16: Orbital frequency as a function of M and R. Detectors need to be sensitive to disturbances at those frequencies since the radiation occurs at the orbital frequency.

is reproduced here:

$$\frac{dP}{dr} = -\frac{G\rho(r)M(r)}{r^2} \quad (8.15)$$

$$\frac{d\left(\frac{P}{\rho_o c^2}\right)}{dr} = -\frac{r_s r}{R^3}$$

The pressure is an energy density, so that $P/\rho_o c^2$ is dimensionless. It is easily solved for

$$r_s = \frac{2GM}{c^2}$$

$$\frac{P(r)}{\rho_o c^2} = \frac{2GM\left(\frac{1}{R} - \frac{r^2}{R^3}\right)}{4} \quad (8.16)$$

$$= \frac{1}{4}\left(\frac{r_s}{R} - \frac{r_s r^2}{R^3}\right)$$

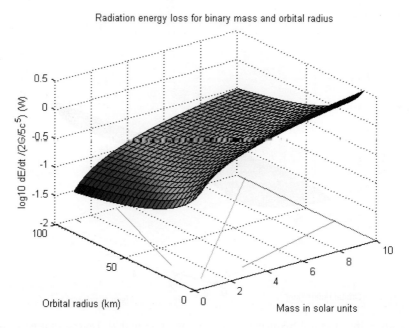

Figure 8.17: Radiated power in watts scaled by the maximum power possible as a function of M and R. Only extreme binary systems with large M and small R approach the maximum.

The pressure is largest at the origin, $P(0)/\rho_o c^2 = r_s/4R$, and disappears at the surface by definition from the boundary conditions.

In the general relativistic case, the tensor for dust without velocity is assumed. In that case a closed form solution is possible which approaches the classical solution when r_s/R approaches zero. The equation for pressure in a Schwarzschild interior space is:

$$P' = \frac{P}{\rho_o c^2}$$

$$\frac{d(P')}{dr} = -\frac{\left(\frac{r_s}{R}\right) * (1 + P') * (1 + 3P') \left(\frac{r}{R^2}\right)}{\left(1 - \frac{r_s r^2}{R^3}\right)} \quad (8.17)$$

which is the Oppenheimer–Volkoff–Tolman equation for a uniform density sphere. There are two new pressure terms and a correction term in the denominator, but in the limit of small Schwarzschild radius and low pressure Eq. (8.17) yields the Newtonian equation.

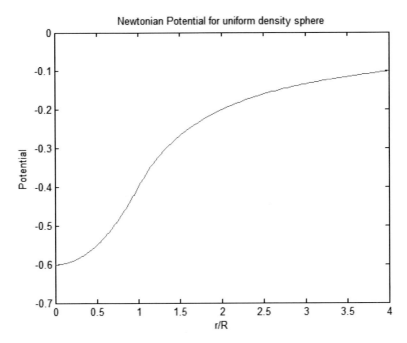

Figure 8.18: Newtonian potential energy for a uniform sphere of radius R. For $r > R$ the potential scales as $1/r$. Arbitrary units, $GM = 1$, are used.

The pressure values are compared in a script called "GR_P_Interior2". The Oppenheimer–Volkoff–Tolman equation is solved symbolically using "dsolve" and the boundary condition $P(R)$ equals zero is imposed.

Printout from that script is shown in Figure 8.19. The symbolic solution for pressure is part of the printout, and the pressure at r equals to zero for classical and GR pressure are compared.

The pressure as a function of radius is shown in Figure 8.20. After some manipulation, the solution of the equation shown in Figure 8.18 can be put into the following form.

$$\frac{P(r)}{\rho_0 c^2} = \frac{\sqrt{\frac{1-r_s}{R}} - \sqrt{\frac{1-r_s r^2}{R^3}}}{\sqrt{\frac{1-r_s r^2}{R^3}} - 3\sqrt{\frac{1-r_s}{R}}} \qquad (8.18)$$

```
>> GR_P_Interior2
   Program to look at Newtonian Potential and Pressure
   and GR Pressure for uniform sphere


Enter rs/R ratio: 0.5
Newtonian Solutions Inside and Outside r = R = 1
Max Newtonian Pressure P(0)/rho*c^2 = rs/4R   =   0.125
Solve GR pressure, dP/dr=-4*pi*G(rho+P/c^2)(rho/3+P/c^2)*r^2/(r-rs
)
                            2
  - ----------------------------------------------- - 1
      /                                         \
      |    d log(b t  - 1) - d log(b - 1)       |
   exp| -  ----------------------------------   | - 3
      \                  3 b                    /
Max GR Pressure P(0)/rho*c^2 = 0.261204
```

Figure 8.19: Printout for the script for a user supplied ratio of r_s/R. The symbolic solution of the equation is also printed out.

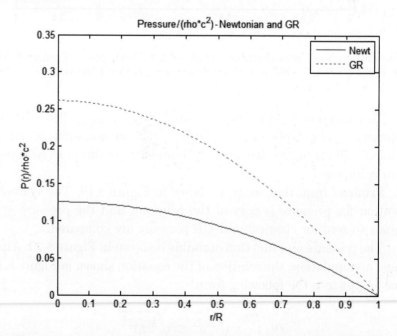

Figure 8.20: Pressures as a function of r for the ratio $r_s/R = 0.5$. The GR contribution is substantial compared to a Newtonian estimate for pressure.

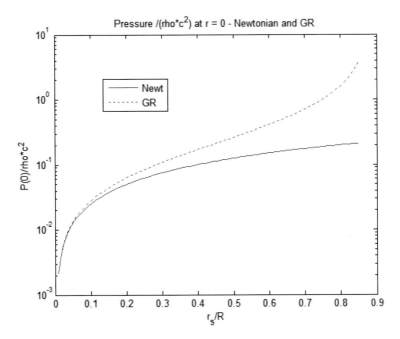

Figure 8.21: Pressures at $r = 0$ as a function of the ratio r_s/R. At low values, GR does not contribute much, but at high values it dominates over the Newtonian estimate.

The maximum pressure, at r equals zero, as a function of the radius R in r_s units is shown in Figure 8.21. As expected, the pressure contributed little at small values. However, at large values the pressure term actually dominates. Clearly, star models for very dense stars must take this into account. In fact, as discussed above, the uniform density approximation badly underestimates the central density and pressure, so that a more realistic model would be a polytropic extension using the Oppenheimer–Volkov–Tolman equation.

The relativistic pressure can be much larger than the pressure exerted by the matter density because pressure acts as a source of gravity, recalling that pressure is an energy density and that all energy has mass and therefore gravitates. Parenthetically, in the collapse of a star of radius R and mass M to a singularity, the change in binding energy is approximately the energy release and is a substantial fraction of the rest energy of the star $GM^2/R \to GM^2/r_s = Mc^2/2$. Hence a supernova collapse releases a truly amazing amount of energy as opposed to fission or fusion which release only a small fraction of the rest mass of their reaction products.

8.8. Light Propagation in the Universe

Since the Universe is expanding, the transmission of light signals is affected because of evolution during the transit time. The expansion means that the apparent velocity of recession, v is roughly proportional to the distance L, with proportionality defined to be the present Hubble constant H_o.

$$v = H_o L \tag{8.19}$$

The wavelength at emission relative to the wavelength at reception is the ratio of the size parameters, R since distances scale with R. The difference in wavelengths defines the z parameter which can be measured by looking at redshifts. Emission is at the time t with size R and reception is indicated by the o subscript. The positive z value means the light is redshifted. For example, the cosmic microwave background has a z value of about 1000. Beyond those z values, earlier in t, electromagnetic information is not available because the universe is opaque. At present size scale at emission, z is zero while at very small size scale at emission, z becomes very large.

$$1 + z = \frac{\lambda_o}{\lambda} = \frac{R_o}{R} \tag{8.20}$$

The Hubble parameter is defined by the expansion rate, set by the rate of change of the scale parameter R, so that it has the dimensions of inverse time.

$$H = \frac{\left(\frac{dR}{dt}\right)}{R} \tag{8.21}$$

The Hubble length is defined by the distance where the recession velocity is c. Any sources beyond that distance are unobservable. The emission time when that occurs is the inverse of the present

```
>> Cosmo_Light
   Program to look at light emission and reception in an expanding cosmology


Enter present Ho in km/sec*million lyr (~30): 30
  Enter estimate for h (~0.73): 0.75
Hubble Length in blyr = 13.3333
Hubble Time in byr = 13.3333
Hubble velocity in c units = 1.5
Present time in byr = 8.88889
```

Figure 8.22: Printout showing the Hubble parameters for a given choice of present Hubble parameter.

Hubble parameter. A matter dominated Universe with an R scaling as a power of t, in this case $n = 2/3$, is assumed in the script "Cosmo_Light". Printout from that script appears in Figure 8.22.

The photons always travel on a path where the local velocity is that of light. Then the present reception and emission radial coordinates, r, are:

$$ds^2 = (c\,dt)^2 - (R\,dr)^2 = 0$$

$$r = c \int_t^{t_o} \frac{dt}{R}$$
(8.22)

The reception time is t_o, while the emission time is t. Assuming a power law behavior in t for R over the past history, the integrals can be performed and the coordinates at emission and reception can be calculated as a function of the z parameter (redshift).

$$r = \frac{c}{R_o} \int_t^{t_o} \frac{dt}{\left(\frac{t}{t_o}\right)^n}$$

$$(1+z) = \left(\frac{t_o}{t}\right)^n = \frac{R_o}{R}$$

$$R_o r = \frac{(ct_o)}{(1-n)} \left[1 - \left(\frac{1}{1+z}\right)^\alpha\right], \quad \alpha = \frac{(1-n)}{n}$$

$$Rr = R_o r \left(\frac{1}{1+z}\right)$$
(8.23)

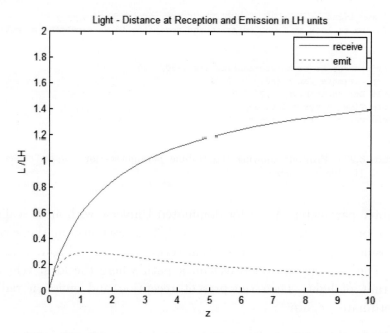

Figure 8.23: The distance at emission and reception for light with redshift z. The distances are scaled to the Hubble length where the recession velocity is c. The present distance at reception is $R_o r$, while the distance at emission is Rr.

The present distance to the source is $R_o r$, while the distance to the source at emission is Rr. There is a maximum present distance of $3ct_o$ which occurs at large z. The emission distance does not increase monotonically with z, but has a maximum. There is a maximum distance that can be observed because of the expansion of the reception point. Note that light can also be received at present distances greater than the Hubble length, also due to the expansion. The Hubble length and velocity are both divided by n in the expanding geometry. The Hubble length is ct/n. The results for a matter dominated Universe, $n = 2/3$, are shown in Figure 8.23. Present distances larger than the Hubble length are accessible.

The look back times for reception and the emission times are shown in Figure 8.24 as a function, again, of the redshift parameter, z. These plots are, in principle, nothing new since the look back positions are already known as shown in Figure 8.23.

8. General Relativity

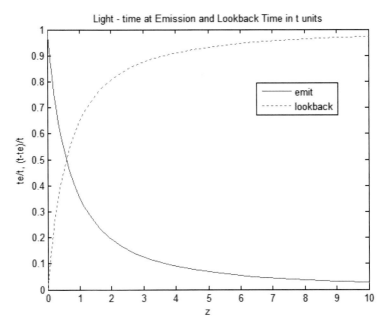

Figure 8.24: Emission time and look back time as a function of the signal redshift. At high z the emission time decreases steadily to zero while the look back time increases toward one.

However, they give additional visual insights. The emission time is the reception time at $z=0$ by definition. The emission time falls steadily as the reception time increases because light arrives from earlier times. The look back time is then just the time of reception with respect to the time of reception and increases monotonically.

Appendix

Scripts for the Chapter on Classical Mechanics

All the scripts are available to the user using the enclosed media. However, it is useful to be able to quickly jump to a written version in order to see what MATLAB commands are used. To that end, the script text for the chapter on Classical Mechanics is enclosed below.

A.1. Angular Momentum

```
%
% Symbolic Angular Momentum
%
clear all;
help Angular_Momentum        % Clear the memory
                               and print header
%
syms x  y  Lz  t  vx  vy
% define variables as needed
%
fprintf('Find L for Symbolic x(t), y(t) Input \n')
%
iloop = 0;
irun = 1;
while irun > 0
    %
    krun = menu('Another x(t), y(t) Trajectory','Yes',
      'No');
```

```
if krun == 2
    irun = -1;
    break
end
%
if krun == 1
    %
    fprintf('Enter x(t) symbolic \n');
    x = input(': ','s'); % Read input as a text
      string
    fprintf('Enter y(t) symbolic \n');
    y = input(': ','s');
    %
    m = 1;
    vx = diff(x,sym('t'));
    vy = diff(y,sym('t'));
    Lz = x *vy - y *vx;
    fprintf('Angular momentum (z-component) = \n');
    Lz
    %
    tt = linspace(0,5);
    for i = 1:length(tt)
       t = tt(i);
        xx(i) = eval(x);
        yy(i) = eval(y);
        LLz(i) = eval(Lz);
        z1(i) = 0;
        x1(i) = 0;
        y1(i) = 0;
    end
    %
    iloop = iloop + 1;
    figure(iloop)
    xmax = max(xx);
    ymax = max(yy);
    zmax = max(LLz);
```

```
        xmin = min(xx);
        ymin = min(yy);
        zmin = min(LLz);
        for i = 1:length(tt)
            %
            plot3(xx(i), yy(i), z1(i),'o');
            axis([xmin xmax ymin ymax zmin zmax])
            %
            title('x,y and Lz ')
            xlabel('x')
            ylabel('y')
            zlabel('Lz')
            pause(0.2)
            hold on
            plot3(x1(i),y1(i),LLz(i),'r*')
            legend('x,y','Lz')
        end
        hold off
        %
    end
end
%
```

A.2. Classical Scattering

```
%
% Program to explore classical scattering
% where in 3 cases the cross section is the same
%
clear all; help Classical_Scatt; % Clear memory and print header
%
fprintf('All Three Scatterings have the Same Total Cross Section \n')
%
iloop = 0;
        %
```

```
b = linspace(0,1.0);     % impact parameter
a = 1; % all results have sigma = pi*a^2
%
% hard sphere , radius a
% barrier, Vo with E = 2Vo
% n = sqrt(1-Vo ./E), bo = n/a
n = sqrt(1- 0.5);
bo = n .*a;
nw = sqrt(1 + 0.5);
%
for i = 1:length(b)
    th_hs(i) = 0;
    if b(i) < a
        th_hs(i) = pi - 2.0 .*asin(b(i) ./a);
    end
    %
    th_bar(i) = 0;
    if b(i) < bo
        th_bar(i) = 2.0 .*(asin(b(i) ./(n .*a))
            - asin(b(i) ./a));
    end
    if b(i) > bo && b(i) < a
        th_bar(i) = pi - 2.0 .*asin(b(i) ./a);
    end
    %
    th_well(i) = 0;
     if b(i) < a
         th_well(i) = -2.0 .*(asin(b(i)
             ./(nw .*a)) - asin(b(i) ./a));
      end
     %
end
%
iloop = iloop + 1;
figure(iloop)
plot(b,th_hs,'-b',b,th_bar,':r',b,th_well,'-.g')
```

```
title('Scattering Angle for Three Scattering
  Centers')
xlabel('b/a')
ylabel('\theta')
legend('Hard Sphere', 'Barrier','Well')
%
```

A.3. Impact Parameter and DCA

```
%
% Program to look at the relationship of impact parameter
% and distance of closest approach - force power laws
%
clear all;              % Clear memory
help CM_b_DCA;          % Print header
%
irun = 1;
iloop = 0;
%
fprintf('Particle with impact parameter b scatters off a
  force center \n');
%
syms b a v vo m r alf aa
%
% central forces, L conserved, mo*vo*b = m * v * a
%
fprintf('a = vo*b/v, a is DCA, initial velocity vo, v at
  DCA \n')
%
fprintf('f(r) = m /r^a, V(r) = -m/(a-1)r^a-1 \n') % force
  and potential
%
fprintf(' E conservation; vo = sqrt(v ^2 + 2 /((alf-1)
  *(r ^(alf-1)))) \n')
%
while irun > 0
    kk = menu(' Another value of the power law a ? ',
      'Yes','No');
```

```
if kk == 2
    irun = -1;
    break
end
if kk == 1
    %
    ALF = input('Enter force law power: ');
    syms b a v vo m r alf aa
      ll = menu('Force Law ? ','attractive',
        'repulsive');
      if ll == 1
          mm = -1;
      else
          mm = 1;
      end
    %
    alf = ALF ;
    aa = solve(a == sqrt(b^2 + mm * (2 *a ^2)
     /(vo ^2*(alf - 1) * (a ^(alf - 1)))),a)
    %
    b = 1;
    vvo = linspace(0.1, 5);
    for i = 1:length(vvo)
        vo = vvo(i);
        AAA(i) = eval(aa(1));
    end
    %
    iloop = iloop + 1;
    figure(iloop)
    loglog(vvo,AAA)
    title('DCA as a function of vo')
    xlabel('vo')
    ylabel('DCA/b')
    %
end
end
%
```

A.4. Foucault Pendulum

```
%
% Program to compute precession of a Focult pendulum
%
clear;
help foucault_exact;
%
% Clear memory and print header
%
syms X1 X2 x1 x2 k1 k2 ;     % k1 is natural frequency =
                                 w^2, k2 is 2*omega*sin
                                 (phi)
                             % phi is latitude, omega is
                                 Earth frequency
                             % pick time in days, omega
                                 = 2*pi
                             % pendulum precesses by
                                 2*pi*sin(phi) per day
                             % x is East and y is North
                                 in Earth system
%
fprintf('Focault pendulum, Gravity plus Coriolis Causes
  Precession \n')
fprintf('D2x = -w^2x + (2*omega*sin(phi))*dydt \n')
fprintf('D2y = -w^2y - (2*omega*sin(phi))*dxdt \n')
%
[X1,X2]=dsolve('D2x1=-k1*x1+k2*D1x2','D2x2=-k1*x2-k2
  *D1x1','x1(0)=0','x2(0)=1','Dx1(0)=0','Dx2(0)=1');
%
iloop = 0;
irun = 1;
while irun > 0
    %
    krun = menu('Another Set of Parameters to Solve?',
      'Yes','No');
```

```
        if krun == 2
            irun = -1;
            break
        end
%
    if krun == 1
        Tpen = input('Enter Pendulum Period (days): ');
         w = (2.0 .*pi) ./Tpen;        % convert to frequency
        phi = input('Enter Latitude (degrees): ');
        phir = (2 .*pi .*phi) ./360; % radians
        omega = 2.0 .*pi; % Earth rotation
        %
        k1 = w .^2;
        k2 = 2.0 .*omega .*sin(phir);
        %
        tt = linspace(0,2,200);
        ntt = length(tt);
        for i = 1:ntt
            t = tt(i);
            xxx1(i) = real(eval(X1));
            xxx2(i) = real(eval(X2));
        end
        %
        iloop = iloop + 1;
        figure(iloop)
        %
        plot(tt,xxx1,'b-',tt,xxx2,'r-')
        title('Exact x(t) and y(t) of Focault Pendulum -
         Earth Frame ')
        xlabel('t')
        ylabel('x, y')
        %
        iloop = iloop + 1;
        figure(iloop)
        plot(xxx1,xxx2,'-b')
        title('Exact x vs y of Focault Pendulum -
         Earth Frame ')
```

```
axis([-1 1 -1 1])
xlabel('x')
ylabel('y')
%
% approximate result
%
ta = linspace(0, tt(ntt),ntt);
%
% in absolute frame, pendulum has SHO,
%
yaf = sin(w .*ta);
%
% rotating - Earth - frame
%
wpre = omega .*sin(phir);
cp = cos(wpre .*ta);
sp = sin(wpre .*ta);
xear = sp .*yaf;
year = cp .*yaf;
%
iloop = iloop + 1;
figure(iloop)
for i = 2:length(ta)
    %
    x1(1) = xear(i-1);
    y1(1) = year(i-1);
    x1(2) = xear(i);
    y1(2) = year(i);
    plot(x1,y1,'-')
    axis([-1 1 -1 1])
    title('Approximate Earth Frame - Precession,
     Movie')
    xlabel('x')
    ylabel('y')
    pause(0.2)
    hold on
end
```

```
      hold off
      %
      iloop = iloop + 1;
      figure(iloop)
      for i = 2:length(ta)
          %
          x1(1) = xxx1(i-1);
          y1(1) = xxx2(i-1);
          z1(1) = 0;
          x1(2) = xxx1(i);
          y1(2) = xxx2(i);
          z1(2) = 0;
          plot3(x1,y1,z1,'-')
          grid ON
          axis([-1 1 -1 1 -1 4])
          title('Exact Earth Frame - Precession and
           Absolute Frame, Movie')
          xlabel('x')
          ylabel('y')
          pause(0.2)
          hold on
          plot3(0,yaf(i),3,'r*')
      end
      hold off
      %
   end
end
%
```

A.5. Spring Pendulum

```
function Spring_Pendulum
%
% Pendulum in plane with length variable due to spring
  connection
%
```

```
global g k m Lo
%
irun = 1;
iloop = 0;
%
while irun > 0
    kk = menu('Pick Another Spring Pendulum?','Yes',
     'No');
    if kk == 2
        irun = -1;
        break
    end
    if kk == 1
        %
        BB = input('Enter k, m and Lo - [k,m,Lo]: ');
        fprintf('Spring contributes as (Lo-r), >0 and
         <0 \n')
        %
        g = 9.8; % m/sec^2
        k = BB(1);
        m = BB(2);
        Lo = BB(3);
        %
        wspr = sqrt(k ./m); % for SHM, period = 2*pi/w
        wpen = sqrt(g ./Lo); % period for simple
         pendullum
        fprintf('w for simple spring = %g and simple
         pendulum = %g \n' , wspr, wpen)
        %
        AA = input('Enter Initial Positions , r and
         theta(deg) - [r(0),theta(0)]: ');
        fprintf('Initial Velocities Assumed to be Zero
         \n')
        %
        ro = AA(1);
        tho = (AA(2) .*2 .*pi) ./360.0;
        %
```

```
tspan = linspace(0,4 .*2.0 .*pi ./wpen);
 % time frame - 4 pendulum periods
%
[t,y] = ode45(@pend_spring,tspan,[0 ro 0 tho]);
 % vr, r vth, th - initial conditions
%
iloop = iloop + 1;
figure(iloop)
plot(t,y(:,2))
title('r vs t')
xlabel('t')
ylabel('r')
%
iloop = iloop + 1;
figure(iloop)
plot(t,y(:,4))
title('\theta vs t')
xlabel('t')
ylabel('\theta')
%
for i = 1:length(t)
    yyy(i) = -y(i,2) .*cos(y(i,4));
    xxx(i) =  y(i,2) .*sin(y(i,4));
end
iloop = iloop + 1;
figure(iloop)
xmax = max(xxx);
xmin = min(xxx);
ymax = max(yyy);
ymax = 0;
ymin = min(yyy);
%
% movie
%
for i = 1:length(t)
    plot(xxx(i),yyy(i),'o')
```

```
                title('Spring - Pendulum Movie - x vs y')
                xlabel('x')
                ylabel('y')
                axis([xmin, xmax, ymin, ymax])
                hold on
                penx(1) = 0.0;
                peny(1) = 0.0;
                penx(2) = xxx(i);
                peny(2) = yyy(i);
                plot(penx,peny,'g')
                plot(0.0,0.0,'r*')
                pause(0.1)
                hold off
            end
            %
            iloop = iloop + 1;
            figure(iloop)
            plot(xxx,yyy,'-')
            title('Spring - Pendulum - x vs y')
            xlabel('x')
            ylabel('y')
            axis([xmin, xmax, ymin, ymax])
            %
        end
end
%--------------------------------------------------------
function dy = pend_spring(t,y)
%
global g k m Lo
dy = zeros(4,1);
% y1 = drdt, 2 = r, 3 = dthetadt, 4 = theta
%
ct = cos(y(4));
st = sin(y(4));
dy(1) = g .*ct + (k ./m) .*(Lo - y(2)) + y(2) .*y(3)
        .*y(3);
```

```
dy(3) = -(g .*st) ./y(2) - (2.0 .*y(1) .*y(3)) ./y(2);
dy(2) = y(1);
dy(4) = y(3);
%
```

A.6. Spherical Pendulum

```
%
% Solve for motion of a spherical pendulum
%
function Spherical_Pend
clear all;
help Spherical_Pend        % Clear the memory and print
                                       header
global Q w
%
fprintf(' Spherical Pendulum, r = 1, m = 1 \n');
fprintf(' Azimuthal Angular Momentum is Conserved -
 1-d Problem \n');
%
irun = 1;
iloop = 0;
while irun > 0
    %
    krun = menu('Another Pendulum IC','Yes','No');
    if krun == 2
        irun = -1;
        break
    end
    if krun == 1;
        %
        ww = input('Enter g/l for the pendulum: ');
        w = sqrt(ww);    % SHO
        %
        Q = input('Enter angular momentum, L/mr^2 : ');
        tho = input('Enter initial angle (deg) : ');
```

```
tho = (2.0 .*pi .*tho) ./360.0 ;
vtho = input('Enter initial angular velocity
  (deg/sec) : ');
vtho = (2.0 .*pi .*vtho) ./360.0 ;
%
tspan = linspace(0, (2.0 .*2.0 .*pi) ./w);
 % time frame - 4 pendulum periods
%
[t,y] = ode45(@sph_pend,tspan,[vtho tho 0]);
 % y1 = dthdt, 2 = th, 3 =
phi, initial conditions
%
iloop = iloop + 1;
figure(iloop)
plot(t,y(:,2))
title('theta vs t')
xlabel('t')
ylabel('\theta')
%
iloop = iloop + 1;
figure(iloop)
plot(t,y(:,3))
title('phi vs t')
xlabel('t')
ylabel('\phi')
%
iloop = iloop + 1;
figure(iloop)
for i = 1:length(y(:,2))
     xx(i) = sin(y(i,2)) .*cos(y(i,3));
     yy(i) = sin(y(i,2)) .*sin(y(i,3));
end
xmax= max(xx);
xmin = min(xx);
ymax = max(yy);
ymin = min(yy);
```

```
            for i = 1:length(y(:,2))
                plot(xx(i),yy(i),'o')
                title('Movie for Spherical Pendulum')
                xlabel('x')
                ylabel('y')
                axis([xmin xmax ymin ymax])
                pause(0.2)
            end
            iloop = iloop + 1;
            figure(iloop)
            plot(xx,yy)
            title('x vs. y')
            xlabel('x')
            ylabel('y')
        end
end
end
%
% ----------------------------------------------------
%
function dy = sph_pend(t,y)
%
global Q w
dy = zeros(3,1);
%
% y1 = dthdt, 2 = th , 3 = phi
%
dy(1) = -w .*sin(y(2)) + (Q .^2 .*cos(y(2)))  
  ./(sin(y(2)) .^3);
dy(2) = y(1);
dy(3) = Q ./(sin(y(2)) .^2);
%
end
%
```

A.7. Driven Pendulum

```
%
% Program to compute a single pendulum, arbitrary angle,
  driven
% using MATLAB tools
%
function Driven_Pendul
%
clear all;
help Driven_Pendul       % Clear the memory and print
 header
%
global vv Adr wdr
%
fprintf('Pendulum - Large Oscillations, Driven,
 1-d Angle \n');
%
irun = 1;
iloop = 0;
%
while irun > 0
    kk = menu('Pick Another Driven Pendulum?',
      'Yes','No');
    if kk == 2
        irun = -1;
        break
    end
    if kk == 1
        %
        % Set initial position and velocity of pendulum
        %
        tho = input('Enter Initial Angle (degrees): ');
        tho = (tho .*pi) ./180.0;    % Convert angle to
          radians
        vo = input('Enter Initial Angular Velocity
          (degrees/sec): ');
```

```
vo = (vo .*pi) ./180.0;
vv = input('Enter Damping in rad/sec: ');
fprintf('Take k/m = 1 - natural frequency units
 \n')
Adr = input('Enter Driving Amplitue: ');
wdr = input('Enter Driving Frequency: ');
%
tt = linspace(0, 12.0 .*pi);
%
% numerical solution using ODE tools
%
[t,y] = ode45(@pend_dr,tt,[vo tho]);
%
iloop = iloop + 1;
figure(iloop)
yy = y(:,1);
plot(t,yy,'-')
title('Angular Velocity')
xlabel('t(sec)')
ylabel('d\theta/dt')
%
iloop = iloop + 1;
figure(iloop)
yy = y(:,2);
plot(t,yy,'-')
title('Angular Position')
xlabel('t(sec)')
ylabel('\theta')
%
iloop= iloop + 1;
figure(iloop)
th = y(:,2);
for j = 1:length(t)
    %
    yy(j) = -cos(y(j,2));   xx(j) = sin(y(j,2));
end
```

```
            xmax = max(xx);   ymax = max(yy);   xmin = min(xx);
             ymin = min(yy);
            for j = 1:length(t)
                xL(1) = 0; yL(1) = 0; xL(2) = xx(j) ; yL(2)
                 = yy(j);
                plot(xx(j),yy(j),'o',xL,yL,'-g')
                hold on
                plot(0,0,'r*')
                axis([xmin xmax ymin ymax])
                title('Movie of Angular Position')
                xlabel('x')
                ylabel('y')
                pause(0.1)
                hold off
            end
            %
        end
end
%
function dy = pend_dr(t,y)
%
global vv Adr wdr
%
dy = zeros(2,1);
dy(1) = -sin(y(2)) - vv .*y(1) + Adr .*cos(wdr .*t);
dy(2) = y(1);
```

A.8. Random Walk

```
%
% Look at a random walk in 2-d
%
clear all;
help rand_walk      % Clear the memory and print header
%
fprintf('Look at a Random Walk in 2-d n')
%
```

```
irun = 1;
iloop = 0;
%
while irun > 0
    kk = menu('Pick Another Number of Random Steps?',...
      'Yes','No');
    if kk == 2
        irun = -1;
        break
    end
    if kk == 1
        %
        Ns = input('Enter Number of Steps of Size 1: ');
        %
        x(1) = 0;
        y(1) = 0;
        NN(1) = 1;
        for i = 2:Ns
            phi = 2 .*pi .*rand;
            x(i) = x(i-1) + cos(phi);
            y(i) = y(i-1) + sin(phi);
            NN(i) = i;
        end
        iloop = iloop + 1;
        figure(iloop)
        %
         xmax = max(x);
         xmin = min(x);
         ymax = max(y);
         ymin = min(y);
         %
         % the movie first, to see how the stochastic
            distance evolves
         %
         for i = 1:Ns
            plot(x(i),y(i),'o')
```

```
            hold on
            title('Random Walk Steps')
            xlabel('x')
            ylabel('y')
            axis([xmin xmax ymin ymax])
            pause(0.1)
        end
        hold off
        %
        iloop = iloop + 1;
        figure(iloop)
        plot(x,y,'-')
        title('Random Walk, Full Time Range')
        xlabel('x')
        ylabel('y')
        axis([xmin xmax ymin ymax])
        %
        iloop = iloop + 1;
        figure(iloop)
        plot(sqrt(NN), sqrt(x .^2+y .^2),'-*')
        title('Distance of Walk vs Number of Steps')
        xlabel('sqrt(N)')
        ylabel('radius')
        %
    end
%
end
%
```

A.9. Rotating Hoop

```
%
% Hoop, radius = 1, Rotates with frequency w with mass
  point located at
% theta, wrt vertical
%
```

```
function Rotating_Hoop2
%
clear all;
help Rotating_Hoop2      % Clear the memory and print header
%
% Initialize
%
global gg ww tho
fprintf('Hoop Rotates with frequency w, w^2 > g/a for equilibrium not equal to 0 \n ')
%
gg = 9.8;    % a = 1 m, pick w
aa = 1;
%
irun = 1;
iloop = 0;
%
while irun > 0
    %
    krun = menu('Another Frequency','Yes','No');

    if krun == 2
        irun = -1;
        break
    end
    if krun == 1;
        %
        ww = input('Enter Frequency of Oscillation,
         > sqrt(g) (~4): ');
        tho = acos(gg /(ww*ww)); % equilibrium angle for
         this w - rad
        fprintf('Stable Angle for Small Oscillations =
         %g (rad)\n' ,tho');
        %
        ro = input('Enter Initial Angle w.r.t. Stable
         Angle (rad): ');
```

```
vo = input('Enter Initial Velocity (rad/sec):');
%
tspan = linspace(0, 5);
[t,y] = ode45(@Hoop,tspan,[vo ; ro]);
%
iloop = iloop + 1;
figure(iloop)
ymax = max(y(:,2));
ymin = min(y(:,2));
for i = 1:length(tspan)
    plot(t(i),y(i,2),'*r')
    title('Angle of Mass Point - Released Near
     Equilibrium Point')
    xlabel('t')
    ylabel('\theta - \theta_o')
    axis([0 max(tspan) ymin ymax])
    pause(0.1)
end
%
iloop = iloop + 1;
figure(iloop)
plot(t,y(:,2),'-b')
title('Angle of Mass Point - Released at
 Equilibrium Point')
xlabel('t')
ylabel('\theta')
%
iloop = iloop + 1;
figure(iloop)
rr = linspace(-1, 1);
zz = sqrt(1- rr .^2);
%
% mass point
%
for i = 1:length(tspan)
    zzm(i) = cos(tho+y(i,2));
```

```
                rrm(i) = sin(tho +y(i,2));
                xxx(i) = rrm(i) .*cos(ww .*t(i));
                rrc = rr .*cos(ww .*t(i));
                plot(rrc,zz,'g',rrc , -zz,'-g',xxx(i),
                  -zzm(i) ,'*r')
                hold on
                plot(sin(tho) .*cos(ww .*t(i)),
                  -cos(tho),'o')
                title('Hoop Rotation, w, and Mass Point
                  Oscillatiion')
                xlabel('x')
                ylabel('y')
                axis([-1.5 1.5 -1.5 1.5])
                hold off
                pause(0.1)
            end
            hold off
            %
        end
end
%
%-------------------------------------------------------------
%
function dydt = Hoop(t,y)
% Equation of mass point on the hoop
%
global gg ww tho
dydt = zeros(2,1);
dydt = [-sin(tho+y(2)).*(gg-ww .*ww .*cos(tho+y(2))) ;
 y(1)];
%
```

A.10. Tides

```
%
% Program to display tides
%
```

```
clear all; help Tides; % Clear memory and print header
%
fprintf('Lunar Tides a la Newton \n')
%
iloop = 0;
%
t = linspace(0,2 .*pi);      % loction of the Moon
te = linspace(0,2 .*pi,20);  % location on earth surface
cth = cos(t);
sth = sin(t);
%
rm = 2;
re = 1;
%
iloop = iloop + 1;
figure(iloop)
%
xm = rm .*cth; % Moon coordinates
ym = rm .*sth;
%
for i = 1:length(t)
    % points for the circle - plots
    x1 = linspace(-re,re);
    y1 = sqrt(1 - x1 .^2);
    hold on
    %
    plot(x1,y1,'r-',x1,-y1,'r-')
    axis([-2 2 -2 2])
    plot(0,0,'g*')
    %
    plot(xm(i), ym(i),'o')
    for j = 1:length(te)
        ang = te(j) - t(i);
        tide(j) = 0.1 .*(3.0 .*cos(ang) .*cos(ang) - 1);
    end
    plot((re+tide) .*cos(te),(re+tide) .*sin(te) ,'b:')
    pause(0.2)
```

```
        hold off
        clf
end
for i = 1:1
        % points for the circle - plots
        x1 = linspace(-re,re);
        y1 = sqrt(1 - x1 .^2);
        hold on
        %
        plot(x1,y1,'r-',x1,-y1,'r-')
        axis([-2 2 -2 2])
        plot(0,0,'g*')
        %
        plot(xm(i), ym(i),'o')
        for j = 1:length(te)
              ang = te(j) - t(i);
              tide(j) = 0.1 .*(3.0 .*cos(ang) .*cos(ang) - 1);
        end
        plot((re+tide) .*cos(te),(re+tide) .*sin(te) ,'b:')
        title('Lunar Tides - Twice per Day')
        pause(0.2)
        hold off
end
        %
```

A.11. Least Action

```
%
% Least Action examples
%
function Lagrange_Act2
%
clear all;
help Lagrange_Act2       % Clear the memory and print header
%
```

```
% Initialize
%
global g delt itype yend1 yend2
%
iloop = 0;
irun = 1;
%
g = 9.8;
%
while irun > 0
    kk = menu('Pick a Least Action','Simple Harmonic',
     'Gravity','Quit');
    if kk == 3
        irun = -1;
        break
    end
    if kk == 1
        %
        itype = kk;
        tf = 2.0;
        vo = -0.1;
        xo = 0;
        fprintf(' SHO - time = 2.0 sec, vo = -0.1 m/sec,
         x(0) = 0 \n ')
        %
        t = linspace(0,tf,20);
        %
        nt = length(t);
        delt = t(2) - t(1);
        %
        % exact trajectory
        %
        ye = xo .*cos(t) + vo .*sin(t); % k = m = 1, scale
         to max amplitude
        yend1 = ye(1);       % fixed end points
        yend2 = ye(nt-1);
        %
```

```
    % closer conversion
    for i = 1:nt
        ysmear(i) = ye(i) + (2.0 .*rand - 1) .* 0.02;
    end
    %
    yysm = fminsearch(@Lla2,ysmear);
    %
    iloop = iloop + 1;
    figure(iloop)
    plot(t,ye,'-',t,yysm,'*r')
    title('Trajectory for Simple Harmonic Motion')
    xlabel('t(sec)')
    ylabel('y(m)')
    legend('Exact','Least Act')
    %
end
    if kk == 2
        %
        itype = kk;
        vo = 10;
        tf = 1;
        fprintf('Gravity - y(0) = 0, t(0) = 0,
         v(0) = 10 m/sec, tf = 1 sec \n')
        %
        t = linspace(0,tf,20);
        %
        nt = length(t);
        delt = t(2) - t(1);
        %
        % exact trajectory
        %
        ye = vo .*t - (g .*t .*t) ./2.0;
        yend1 = ye(1);    % fixed end points
        yend2 = ye(nt-1);
        %
        % closer conversion
```

```
            for i = 1:nt
                ysmear(i) = ye(i) + (2.0 .*rand - 1)
                    .* 0.02;
            end
            yysm = fminsearch(@Lla2,ysmear);
            %
            iloop = iloop + 1;
            figure(iloop)
            plot(t,ye,'-',t,yysm,'*r')
            title('Trajectory in Gravity Field')
            xlabel('t(sec)')
            ylabel('y(m)')
            legend('Exact','Least Act')
            %
        end
end
%
%-----------------------------------------------------
%
function y = Lla2(x)
global g delt itype yend1 yend2
% evaluate action T, V along time
        %
        nt = length(x);
        vla = diff(x);      % velocity, kinetic and
        potential, nt-1
        vla = vla ./delt;
        if itype == 1;
            for i = 1:nt-1
                Tla(i) = (vla(i) .^2) ./2;
                Vla(i) = -x(i) .^2;
            end
        end
        if itype == 2;
            for i = 1:nt-1
                Tla(i) = (vla(i) .^2) ./2;
```

```
              Vla(i) = g .*x(i);
          end
      end
      %
      y = sum(Tla - Vla) .*delt;
      %
      % penalize end points not fixed
      y = y + 1.0 .*(abs(x(1)-yend1) + abs(x(nt-1) -
        yend2));
      %
```

A.12. Rocket Drag

```
%
function cm_rocket_drag
%
% Solve non-relativistic rocket, numerically
% allow for uniform gravity field and air drag
%
clear all;
help cm_rocket_drag     % Clear the memory and print
                          header
%
% solve the rocket equation
%
global vo dmdt mo g dragg zo
%
g = 9.8; % accel at earth surface m/sec^2
re = 6.378 .*10 .^6; % earth radius - m
veq = (2.0 .*pi .*re) ./(24 .*3.6 .*10^3); % equatoria_
  launch velocity km/sec
rs = 1.5 .*10 .^11;  % distance to sun - m
me = 6.0 .*10 .^24;  % earth mass - kg
ms = 2.0 .*10 .^30;  % sun mass, - kg
%
vorb = sqrt(g .*re);   % orbital velocity - circular,
  low orbit
```

```
ve =sqrt(2.0 .*g .*re); % escape velocity for Earth ~
 11.2 km/sec
vs = ve .*sqrt(ms .*re ./(me .*rs)); % escape velocity to
 leave solar system ~ 42 km/sec
%
% air density falloff with altitude - m
zo = 9140;
dragg = 0;
%
fprintf('Velocity, Satellite Low Circular Orbit (m/sec)
 = %g \n',vorb);
fprintf('Escape Velocity - Earth (m/sec) = %g \n',ve);
fprintf('Escape Velocity - Solar System (m/sec) = %g \n',
 vs);
fprintf('Equatorial Launch Velocity (m/sec) = %g \n',
 veq);
%
irun = 1;
iloop = 0;
%
while irun > 0
    kk = menu('Pick Another Drag Coefficient?','Yes',
      'No');
    if kk == 2
        irun = -1;
        break
    end
    if kk == 1
        %
        % total possible burn time T is mo/(dm/dt)
        % payload ratio mp/mo = 1-tp/T, tp = burn time
          for this payload
        %
        mo = input('Input the Rocket Mass (in 10^6 kg
          units) - Saturn = 4x10^6 kg: ');
        mo = mo .*10 .^6;
```

Appendix

```
mp = input('Input the Payload Mass (in kg) -
 Saturn Escape Module = 24610 kg: ');
vo = input('Input the Exhaust Velocity (in m/sec)
 - Saturn = 2200 m/sec: ');
dmdt = input('Input Burn Rate (in kg/sec) -
 Saturn = 15000 kg/sec: ');
%
T = mo ./dmdt ; % max possible burn rate, with no
 payload
tp = T .*(1 - mp ./mo); % burn time for this
 payload
tlift = -vo ./g + T; % t=0 is ignition, t = tl
 is lift time, when acceleration > 0
%
fprintf('Maximum Burn Time (sec) = %g \n',T);
fprintf('Burn Time for This Payload (sec) = %g
 \n',tp);
%
tspan = linspace(tlift,tp,100);
g = 0;
[t,yfree] = ode45(@drag,tspan,[0 0]);
g = 9.8;
[t,yg] = ode45(@drag,tspan,[0 0]);
%
dragg = input('Input Air Drag - Viscosity: ');
[t,ydrag] = ode45(@drag,tspan,[0 0]);
%
iloop = iloop + 1;
figure(iloop)
semilogy(t ./T,yfree(:,2),'-b',t ./T,yg(:,2),
 '-r',t ./T,ydrag(:,2),'-g')
legend('free','g','g+drag')
title('Rocket Motion With Gravity and Air Drag')
xlabel('Fractional Burn Time')
ylabel('y(m)')
%
```

```
            iloop = iloop + 1;
            figure(iloop)
            semilogy(t ./T,yfree(:,1),'-b',t ./T,yg(:,1),
              '-r',t ./T,ydrag(:,1),'-g')
            hold on
            semilogy([-0.4, 0.4], [ve, ve],'-k')
            semilogy([-.4 0.4 ], [vs, vs],':k')
            legend('free','g','g+drag')
            title('Rocket Motion With Gravity and Air Drag')
            xlabel('Fractional Burn Time')
            ylabel('v(m/sec)')
            hold off
      end
end
%
%-----------------------------------------------------
%
function dy = drag(t,y )
%
% rocket in uniform gravity field with air drag -
  vertical motion
%
global vo dmdt mo g dragg zo
%
dy = zeros(2,1);
%
m = mo - dmdt .*t;
dy(1) = (vo .* dmdt) ./m - g - ( dragg .*y(1) .*y(1)
  .* exp(-y(2) ./zo)) ./m;
dy(2) = y(1);     % 2 position, 1 velocity
%
```

A.13. Two Stage Rocket

```
%
% Rockets - 2 stage vs. 1 stage
%
```

```
clear all;
help Two_Stage         % Clear the memory and print header
%
% solve the rocket equation - no gravity
% compare 2 and 1 stage approaches for the same final
  payload
%
gg = 9.8; % accel at earth surface m/sec^2
re = 6.378 .*10 .^6;  % earth radius - m
rs = 1.5 .*10 .^11;   % distance to sun - m
me = 6.0 .*10 .^24;   % earth mass - kg
ms = 2.0 .*10 .^30;   % sun mass, - kg
%
vorb = sqrt(gg .*re);    % orbital velocity - circular,
  low orbit
ve =sqrt(2.0 .*gg .*re); % escape velocity for Earth -
  11.2 km/sec
vs = ve .*sqrt(ms .*re ./(me .*rs)); % escape velocity
  to leave solar system - 42 km/sec
%
fprintf('Velocity, Satellite Low Circular Orbit (m/sec)
  = %g \n',vorb);
fprintf('Escape Velocity - Earth (m/sec) = %g \n',ve);
fprintf('Escape Velocity - Solar System (m/sec) = %g \n',
  vs);
%
irun = 1;
iloop = 0;
%
while irun > 0
    kk = menu('Pick Another Two Rockets?','Yes','No');
    if kk == 2
        irun = -1;
        break
    end
    if kk == 1
        %
```

```
% vf-vi = vo*log(mf/mi) - valid generally
% m(t) = mi - dmdt*(t-ti), assume constant burn
  rate
%
mo = input('Input the Rocket Mass (in 10^6 kg
  units) - Saturn = 4x10^6 kg: ');
mo = mo .*10 .^6;
mp = input('Input the Payload Mass (in kg) -
  Saturn Escape Module = 24610 kg: ');
vo = input('Input the Exhaust Velocity (in m/sec)
  - Saturn = 2200 m/sec: ');
dmdt = input('Input Burn Rate (in kg/sec) -
  Saturn = 15000 kg/sec: ');
m2 = input('Input Mass for Second Stage(in 10^6
  kg units) : ');
m2 = m2 .*10 .^6;
rat = input('Input the Plumbing to Fuel Mass
  Ratio for Both Stages : ');
%
% defines the masses, same payload in both cases
% single stage
%
Mfu = (mo - mp) ./(1.0 + rat); % fuel is reduced
  by plumbing deadweight
vf1 = -log((mp + rat .*Mfu) ./mo);
tf1 = Mfu ./dmdt;
%
% 2 stage assuming same plumbing ratio in both
  cases
%
mfu2 = (m2 - mp) ./(1.0 + rat); % second stage
  mass
mfu1 = (mo - m2) ./(1.0 + rat); % first stage,
  same total mass = mo
mr2 = m2 - mp - mfu2;
mr1 = mo - m2 -mfu1;
```

```
tfu1 = mfu1 ./dmdt;
tfu2 = mfu2 ./dmdt;
vfu1 = -log((m2 + mr1) ./mo);
vfu2 = vfu1 - log((mp + mr2) ./m2);
%
T = mo ./dmdt ; % max possible burn time
%
tt = linspace(0,tf1); % one stage
%
% the free rocket
%
for i = 1:length(tt)
    %
    v1(i) = 0;
    if tt(i) < tf1
        v1(i) = - vo .*log(1.0 - dmdt .*tt(i)
           ./mo);   % 1 stage
    end
    %
end
for i = 1:length(tt)
    v2(i) = 0;
    if tt(i) < tfu1 + tfu2
        if tt(i) < tfu1;      % first stage
            v2(i) = -vo .*log(1.0 - (dmdt
              .*tt(i)) ./mo);   % first stage
        else
            v2(i) = vfu1 .*vo - vo .*log(1.0 -
              (dmdt .*(tt(i)-tfu1)) ./m2);
        end
    end
end
%
iloop = iloop + 1;
figure(iloop)
semilogy(tt,v1 ./vo,'-b',tt, v2./vo,'-g')
```

```
            title(' Rocket - Velocity in vo Units')
            xlabel('Burn Time (sec)')
            ylabel('velocity/vo')
            legend('1 stage','2 stage')
            %
            hold on
            %
            semilogy(tt,vorb./vo,'r-',tt,ve ./vo,'r:',tt,vs
              ./vo,'r--')
            legend('One Stage','Two Stages','Orbital
              Velocity', 'Earth Escape
            Velocity', 'Sun Escape Velocity')
            hold off
            %
            iloop = iloop + 1;
            figure(iloop)
            plot(tt ./T,v1 ./vo,'-b',tt ./T, v2./vo,'r:')
            title(' Rocket - Velocity in vo Units')
            xlabel('Burn Time in T Units')
            ylabel('velocity/vo')
            legend('1 stage','2 stage')
      %
      end
end
%
```

A.14. Table Top

```
%
% Program to look at particles on a plane table,
   (r, theta) and hanging
% below (z) with force g
%
function Table_Top
clear all;               % Clear memory
help Table_Top;          % Print header
%
```

```
global g m1 m2 L
irun = 1;
iloop = 0;
%
while irun > 0
    kk = menu(' Another Set of masses, m1 and m2 ? ',
      'Yes','No');
    if kk == 2
        irun = -1;
        break
    end
    if kk == 1
        %
        m1 = input('Enter Mass on the Table (kg): ');
        a = input('Enter m1 Initial Radius (m): ');
        dadt = input('Enter m1 Initial Radial Velocity
          (m/sec): ');
        m2 = input('Enter Mass Hanging Below the Table
          (kg) at r = 0: ');
        L = input('Enter m1 Angular Momentum (m^2/sec):');
        g = 9.8;     % m/sec^2
        %
        fprintf('Masses Connected by String, lenghth LL,
          z = LL-r \n')
        LL = input('Enter String Length (m): ');
        %
        tspan= linspace(0,10);
        [t,y] = ode45(@top_table,tspan,[dadt a 0]);
         % initial radial velocity and position
        %
        iloop = iloop + 1;
        figure(iloop)
        %
        plot(t,y(:,2))
        title('r vs t')
        xlabel('t')
```

```
ylabel('r')
%
iloop = iloop + 1;
figure(iloop)
plot(t,y(:,1))
title('dr/dt vs t')
xlabel('t')
ylabel('dr/dt')
%
iloop = iloop + 1;
figure(iloop)
plot(t,LL - y(:,2))
title('z of m2 vs t')
xlabel('t')
ylabel('z')
iloop = iloop + 1;
figure(iloop)
plot(t,y(:,3))
title('polar angle vs t')
xlabel('t')
ylabel('\theta')
%
% movie
%
for i = 1:length(t)
    yyy(i) = y(i,2) .*cos(y(i,3));
    xxx(i) = y(i,2) .*sin(y(i,3));
    zzz(i) = 0;
    xx(i) = 0;
    yy(i) = 0;
    zz(i) = -(LL - y(i,2));
end
iloop = iloop + 1;
figure(iloop)
xmax = max(xxx);
xmin = min(xxx);
```

```
ymax = max(yyy);
ymin = min(yyy);
zmax = 0.1;
zmin = - LL;
%
% movie
%
for i = 1:length(t)
    plot3(xxx(i),yyy(i),zzz(i),'bs')
    title('Plane and Hanging Weight')
    xlabel('x')
    ylabel('y')
    zlabel('z')
    axis([xmin, xmax, ymin, ymax, zmin, zmax])
    grid
    hold on
    penx(1) = 0.0;
    peny(1) = 0.0;
    penz(1) = 0.0;
    penx(2) = xxx(i);
    peny(2) = yyy(i);
    penz(2) = zzz(i);
    plot3(penx,peny,penz,'g')
    plot3(0,0,0,'*r')
    plot3(xx(i),yy(i),zz(i),'bs')
    wtx(1) = 0.0;
    wty(1) = 0.0;
    wtz(1) = 0.0;
    wtx(2) = xx(i);
    wty(2) = yy(i);
    wtz(2) = zz(i);
    plot3(wtx,wty,wtz,'g')
    grid
    pause(0.2)
    hold off
    %
```

```
            end
                iloop = iloop + 1;
                figure(iloop)
                plot3(xxx,yyy,zzz,'b-')
                title('Plane and Hanging Weight')
                xlabel('x')
                ylabel('y')
                zlabel('z')
                grid
                axis([xmin, xmax, ymin, ymax, zmin, zmax])
                hold on
                plot3(xx,yy,zz,'b-')
                plot3(0,0,0,'*r')
                hold off
            %
        end
%
    end
%
end
% ----------------------------------------------------------
function dy = top_table(t,y)
%
global g m1 m2 L
dy = zeros(3,1);
%
% y1 = drdt, 2 = r , 3 = th
%
dy(1) = -g .*m2 + (L .^2 .*m1) ./(y(2) .^3);
dy(2) = y(1);
dy(3) = L ./(y(2) .^3);
%
end
%
```

A.15. Eccentric_Kepler

```
%
% Program to compute conic sections with different
  eccentricity
%
clear;
help eccentric_Kepler;
%
% Clear memory and print header
%
fprintf('Keplerian Orbits - defined by Eccentricity \n')
fprintf('r = ro*(1+e)/(1+e*cos(theta)) \n')
%
ro = 1;
th = linspace(0,2 .*pi,100);
st = sin(th);
ct = cos(th);
%
iloop = 1;
irun = 1;
while irun > 0
    %
    krun = menu('Another Eccentricity?','Yes','No');
    if krun == 2
        irun = -1;
        break
    end
    %
    if krun == 1
        %
    ecc = input('Enter Eccentricity, (0,2): ');
    if ecc > 1
        thasy = (360 .*acos( 1.0 ./ecc)) ./(2.0 .*pi);
        fprintf('e > 1 - Aysmptotic Angle = %g \n',
          thasy);
```

```
        end
    r = ro .*(1.0 + ecc) ./(1+ecc .*ct);
    x = r .*ct;
    y = r .*st;
    %
    figure(1)
    %
    plot(x,y,'-',0,0,'r*')
    title('Orbit/Conic Section for All Eccentricities ')
    xlabel('x')
    ylabel('y')
    axis([-10 10 -10 10])
    hold on
    iloop = iloop + 1;
    figure(iloop)
    plot(x,y,'-',0,0,'r*')
    title('Orbit/Conic Section for This Eccentricity ')
    xlabel('x')
    ylabel('y')
%
    end
    hold off
end
%
```

A.16. Non-Central Force

```
function Non_Central_Force
%
% orbits - Kelperian except that L is not conserved -
  driving harmonic
% force
%
global pert w
%
irun = 1; iloop = 0;
%
```

```
while irun > 0
    kk = menu('Pick Another Strength of Perturbation?',
    'Yes','No');
    if kk == 2
        irun = -1;
        break
    end
    if kk == 1
        %
        pert = input('Enter Strength of Perturbation for
          Elliptical Orbit (~0.3):   ');
        w = input('Enter Frequency, w, of Perturbation
          (~ 1):  ');
        a = input('Enter Initial radius (~ 4): ');
        LL = input('Enter Initial Angular Velocity
          1/a^3/2 Circular: ');
        fprintf('Angular velocity < circular is not
          stable and hits r = 0 \n')
        %
        % circular orbits
        %
        Lc = sqrt(a);
        vc = 1.0 ./(a .^1.5); % angular velocity 1/a^2/3
        tauc = 2.0 .*pi .*(a .^1.5);
        %
        fprintf('Circular Orbits for Initial Radius
          a \n')
        fprintf('L = %g, angular velocity = %g, period =
          %g \n',Lc,vc,tauc)
        %
        tspan = linspace(0,4 .*tauc); % time frame - 4
          circular periods
        %
        [t,y] = ode45(@mech,tspan,[0 a LL 0]); % vr, r
          vth, th - initial
          conditions
```

```
%for unperturbed ellipse
%
iloop = iloop + 1;
figure(iloop)
plot(t,y(:,2))
title('r vs t')
xlabel('t')
ylabel('r')
%
iloop = iloop + 1;
figure(iloop)
plot(t,y(:,4))
title('\theta vs t')
xlabel('t')
ylabel('\theta')
%
for i = 1:length(t)
    xxx(i) = y(i,2) .*cos(y(i,4));
    yyy(i) = y(i,2) .*sin(y(i,4));
end
iloop = iloop + 1;
figure(iloop)
xmax = max(xxx);
xmin = min(xxx);
ymax = max(yyy);
ymin = min(yyy);
%
% movie
%
for i = 1:length(t)
    plot(xxx(i),yyy(i),'o')
    title(' Perturbed Orbit Movie - x vs y')
    xlabel('x')
    ylabel('y')
    axis([xmin, xmax, ymin, ymax])
    hold on
```

```
                plot(0.0,0.0,'r*')
                pause(0.1)
            end
            %
        end
end
%------------------------------------------------------------
function dy = mech(t,y)
%
global pert w
dy = zeros(4,1);
% y1 = drdt, 2 = r, 3 = dthetadt, 4 = theta
%
dy(1) = (y(3) .^2) .*y(2) - 1.0 ./y(2) .^2;
dy(3) = (-2 .*y(1) .*y(3) .*y(2) + pert .*sin(w*t))
 ./(y(2) .^2);
dy(2) = y(1);
dy(4) = y(3);
%
```

A.17. Euler_Angles

```
%
% Graphics for Euler Angles
%
clear all;
help Euler_Angles     % Clear the memory and print header
%
% define variables as needed
%
% circle orientiation for coord system
phi = linspace(0,2 .*pi);
x = cos(phi);
y = sin(phi);
nph = length(phi);
for i = 1:nph
```

```
    z(i) = 0;
end
%
% x axis
%
xax(1) = 0;
xax(2) = 1;
yax(1) = 0;
yax(2) = 0;
zax(1) = 0;
zax(2) = 0;
%
% z axis
%
xzx(1) = 0;
xzx(2) = 0;
yzx(1) = 0;
yzx(2) = 0;
zzx(1) = 0;
zzx(2) = 1;
%
% now successive rotations
%
ang = 35 .*2.0 .*pi ./360. ; % all 30 degrees;
sa = sin(ang);
ca = cos(ang);
D = [ca -sa 0; sa ca 0; 0 0 1];
C = [1 0 0; 0 ca sa; 0 -sa ca];
B = [ca sa 0; -sa ca 0; 0 0 1];
%
for i = 1:nph
    xx(1) = x(i);
    xx(2) = y(i);
    xx(3) = z(i);
    xD(:,i) = D*xx';
    xCD(:,i) = C*D*xx';
```

```
    xBCD(:,i) = B*C*D*xx';
end
%
% x axes
%
for i = 1:2
    xxx(1) = xax(i);
    xxx(2) = yax(i);
    xxx(3) = zax(i);
    xaxD(:,i) = D*xxx';
    xaxCD(:,i) = C*D*xxx';
    xaxBCD(:,i) = B*C*D*xxx';
end
%
for i = 1:2
    xxzx(1) = xzx(i);
    xxzx(2) = yzx(i);
    xxzx(3) = zzx(i);
    xzxD(:,i) = D*xxzx';
    xzxCD(:,i) = C*D*xxzx';
    xzxBCD(:,i) = B*C*D*xxzx';
end
%
plot3(x,y,z,'b-');      % initial orientation
hold on
plot3(xax,yax,zax,'b-')
plot3(xD(1,:),xD(2,:),xD(3,:),'-.r')       % rotate on
 (x,y) plane
plot3(xCD(1,:),xCD(2,:),xCD(3,:),'g--')
plot3(xBCD(1,:),xBCD(2,:),xBCD(3,:),':k')
plot3(0,0,0, 'r*')
plot3(xaxD(1,:),xaxD(2,:),xaxD(3,:),'r-.')
plot3(xaxCD(1,:),xaxCD(2,:),xaxCD(3,:),'g--')
plot3(xaxBCD(1,:),xaxBCD(2,:),xaxBCD(3,:),'k:')
plot3(xzx,yzx,zzx,'b-')
plot3(xzxD(1,:),xzxD(2,:),xzxD(3,:),'r-.')
```

```
plot3(xzxCD(1,:),xzxCD(2,:),xzxCD(3,:),'g--')
plot3(xzxBCD(1,:),xzxBCD(2,:),xzxBCD(3,:),'k:')
xlabel('x')
ylabel('y')
zlabel('z')
title('Euler Angles')
axis([-1 1 -1 1 -0.4 0.4])
hold off
%
```

A.18. Top Forceless

```
%
% Top Motion with no forces
%
clear all;
help Top_Forceless      % Clear the memory and print
 header
%
% define variables as needed
%
fprintf('Top Motion Depends on Top moments of I \n')
fprintf('and Angular Velocity w \n')
%
iloop = 0;
irun = 1;
while irun > 0
    %
    krun = menu('Another Set of Top I3, I','Yes','No');
    if krun == 2
        irun = -1;
        break
    end
    %
    if krun == 1
        %
        I1 = 1;
```

Appendix

```
Iz = input('Enter Top Shape w = 1, Iz, I1 = 1
  (~1.5) : ') ; % moments of 2 principal axes
%
tho = input('Enter Initial Top Polar Angle (deg)
  (~20) : ');
tho = (tho .*2 .*pi) ./360;      % theta (0) -
  rad, must be non-zero(sleeping)
%
w = 1;
w3 = w .*cos(tho);
wt = w .*sin(tho);
omeg = ((Iz - I1) .*w3) ./I1 ;
%
tt = linspace(0, 4 .*pi,50);
w1 = wt .*cos(omeg .*tt);
w2 = wt .*sin(omeg .*tt);
for i = 1:length(tt)
    w3t(i) = w3;
end
%
iloop = iloop + 1;
figure(iloop)
xmax = max(w1);
ymax = max(w2);
zmax = max(w3t);
xmin = min(w1);
ymin = min(w2);
zmin = min(w3t);
for i = 1:length(tt)
    x1(1) = 0;
    x1(2) = w1(i);
    y1(1) = 0;
    y1(2) = w2(i);
    z1(1) = 0;
    z1(2) = w3t(i);
    %plot3(w1(i), w2(i), w3t(i),'o');
```

```
                    plot3(x1,y1,z1,'-r*')
                    axis([xmin xmax ymin ymax 0 zmax])
                    title('Top Precession ')
                    xlabel('wx')
                    ylabel('wy')
                    zlabel('wz')
                    pause(0.2)
                    hold on
            end
            hold off
            %
        end
end
```

A.19. Top_Euler

```
%
% Top Motion using explicit solution, Euler Equations
%
function Top_Euler
%
clear all;
help Top_Euler       % Clear the memory and print header
global B Iz I1 ws
%
% define variables as needed
%
fprintf('Top motion depends on spin, Top moments of I,
  mass \n')
fprintf('and initial posiition and velocity (4 of) \n')
fprintf('Special Case of "drop", theta = phi = phidot =
  0, only thetao \n')
%
iloop = 0;
irun = 1;
```

```
while irun > 0
    %
    krun = menu('Another Set of Top Parameters -
    Euler Angle
    Solution?','Yes','No');
    if krun == 2
        irun = -1;
        break
    end
    %
    if krun == 1
        B = input('Enter m*g*h -Torque at \theta
          (~1.0) = 0 : ');
        I1 = 1;
        Iz = input('Enter Top Shape Iz, I1 = 1 (~1.5): ');
        % moments of 2 principal axes
        %
        tho = input('Enter Initial Top Polar Angle (deg)
          (~20) : ');
        tho = (tho .*2 .*pi) ./360;     % theta (0) - rad,
          must be non-zero(sleeping)
        phio = 0;       % phi(0)
        phio = (phio .*2 .*pi) ./360;
        psio = 0;
        thdto = 0; % thetadot(0)
        %
        phidto = input('Enter Initial Derivative of Top
          Azimuthal Angle - Defines
          Nutation (deg) (~0,0.2,0.8) : ');
        %
        ws = 1.5 + sqrt(4 .*I1 .*B .*cos(tho))/Iz ;
        %
        % time scale for precession is ~ 1/sqrt(B),
          integrate eqs for theta(t),
        % phi(t)using Euler Angles in body axes,
          transform to space axes
        %
```

```
range = 1.0 ./sqrt(B);
%
tspan = linspace(0, 8 .*range,200);
% y(1) = phi, y(2)= phid, y(3) = th, y(4) = thd,
 y(5) = psi, y(6) = s
% s is a constant of the motion = top spin,
[tt,yt] = ode45(@top_euler_eqs,tspan,[phio,
 phidto, tho, thdto, psio] );
% constant of the motion
psidot = ws - yt(:,2) .*cos(yt(:,3));
%
iloop = iloop + 1;
figure(iloop)
plot(tt ./range,yt(:,3).*360 ./(2.0 .*pi))
title('\theta - nutation ')
xlabel('time in 1/sqrt(B) units')
ylabel('theta degrees')
%
iloop = iloop + 1;
figure(iloop)
plot(tt ./range,yt(:,1).*360 ./(2.0 .*pi))
title('\phi - precesion ')
xlabel('time in 1/sqrt(B) units')
ylabel('phi degrees')
%
iloop = iloop + 1;
figure(iloop)
plot(yt(:,1).*360 ./(2.0 .*pi) ,yt(:,3).*360
 ./(2.0 .*pi));
title('\phi vs \theta ')
xlabel('\phi (deg)')
ylabel('\theta(deg)')
%
iloop = iloop + 1;
xx = sin(yt(:,3)) .*cos(yt(:,1));
yy = sin(yt(:,3)) .*sin(yt(:,1));
```

```
            zz = cos(yt(:,3));
            xmax = max(xx);
            xmin = min(xx);
            ymax = max(yy);
            ymin = min(yy);
            zmax = max(zz);
            zmin = min(zz);
            for i = 1:length(xx)
            plot3(xx(i),yy(i),zz(i),'r*')
            axis([xmin xmax ymin ymax zmin zmax])
            title('CM Motion of the Top')
            xlabel('x')
            ylabel('y')
            zlabel('z')
            hold on
            pause(0.1)
            end
            hold off
            plot3(xx,yy,zz)
            axis([xmin xmax ymin ymax zmin zmax])
            title('CM Motion of the Top')
            xlabel('x')
            ylabel('y')
            zlabel('z')
      end
end
%
%-----------------------------------------------
%
function dy = top_euler_eqs(t,y)
%
global B Iz I1 ws
dy = zeros(5,1);        % one constant of motion, top spin
%
% y(1) is phi of top, y(2) is phidot , y(3) is th,
 y(4) is thdot, y(5) is psi
%
```

```
dy(1) = y(2);
dy(3) = y(4);
dy(2) = (Iz .*ws - 2.0 .*I1 .*y(2).*cos(y(3))).*y(4)
     ./(I1*sin(y(3)))   ;
dy(4) = (B - (Iz .*ws - I1 .*y(2).*cos(y(3))).*y(2))
     .*sin(y(3))/I1   ;
dy(5) = ws - y(2) *cos(y(3));
%
```

Graphical User Interface (GUI)

All the scripts used in this volume are command line driven in that the user makes input via the command window using text. In fact, MATLAB has a suite of tools allowing the use of GUIs in order to wrap the scripts and have a graphical input of user choices for a specific problem. There are several "help" items in the "product help" giving GUI details. In particular there is a video demo which illustrates how to build a GUI using the MATLAB supplied tools.

When creating a GUI, there is a palette of possible GUI items. The choices are shown below. Having chosen an object, the "property editor" allows the GUI designer to set the properties, for example, the "Tag" to identify the object or the String defining the text on the pushbutton.

The use of a GUI rather than a command line driven input dialogue is illustrated with two examples. These examples are "wrappers" for the script in Section 4.1 (Leaky Box) and 5.13 (Damped, Driven Oscillator), First the problem of gas leaking from a container is wrapped with a GUI which features editable text boxes which receive user input rather than command line input. Default values are given as a text string. The number of molecules, their temperature and the size of the aperture are supplied by the user, or the defaults can be run using the start pushbutton.

The gas problem is one where number, temperature and "volume" for the ideal gas are defined but the "pressure" must be approximately evaluated by running the script. In another problem, the harmonic oscillator with damping being driven harmonically by a wave with specified amplitude and frequency can be done in real time. In that case, sliders are more appropriate. The slider limits

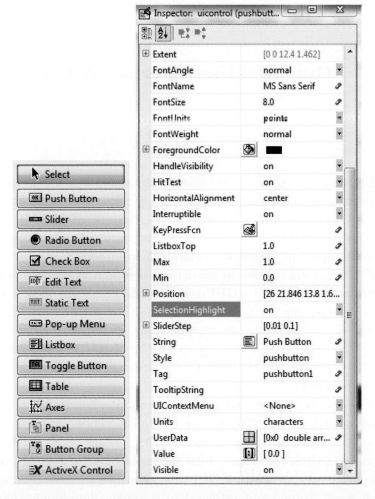

Figure GUI.1: Tools for GUI design in MATLAB. The pallette is on the left and the property editor for a pushbutton is on the right.

are provided in text boxes and the calculation starts and stops with pushbuttons. During the calculation the user can choose the values using the sliders. That implementation makes the response of the oscillator very intuitively realizable. A snapshot of the GUI for this problem is shown below.

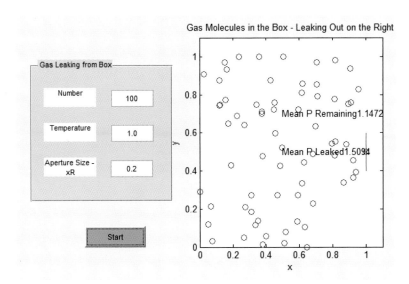

Figure GUI.2: GUI for the leaking box problem showing the use of editable text, text boxes and a pushbutton. The result for the mean momentum of the remaining and escaped gas appears as annotated text on the figure of the box.

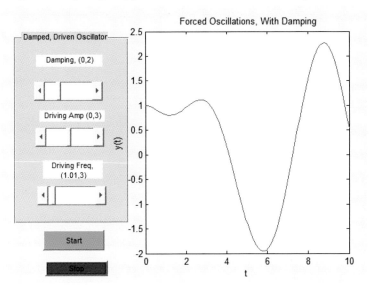

Figure GUI.3: Snapshot of the waveform for a harmonic oscillator with a particular user chosen set of slider parameters for damping and driven amplitude and frequency.

References

Numerical Methods for Physics, 2e, Alejandro L. Garcia, Prentice Hall, 2000, ISBN: 0-13-906744-2. This text has companion script which is very useful as a starting point for some exercises.

Numerical Recipes in Fortran 77. The Art of Scientific Computing, W. H. Press, S. A. Teukolsky, W. T. Vetterling, and B. P. Flannery, Cambridge University Press, 1986, ISBN: 0-521-43064-X. This book is a veritable encyclopedia of numerical methods.

There are many excellent textbooks available for use as references. However, as technology has evolved, the increasing use of online resources is at hand. Therefore, a complete set of paper textbooks is not quoted here. Rather the search engines and compiled online knowledge bases are invoked.

Google is an enormously useful search engine and many specific searches will yield a great variety of information. Wikipedia has a large store of interesting Physics topics, and a search through them will very often start the user on a good path. Indeed, a particular article often has many links that can be followed deeper into the topic. An example of the first page of a search for "Foucault pendulum" is shown below. There are additional links and references provided that give the user a very good reference experience. Indeed, while looking at the MATLAB scripts for this text, the user can easily dip into the online resources and gain further knowledge.

Foucault pendulum

From Wikipedia, the free encyclopedia
(Redirected from Focault pendulum)

This article is about the physics experiment and implement. For the novel by Italian philosopher Umberto Eco, see Foucault's Pendulum.

The **Foucault pendulum** (English pronunciation: /fuːˈkoʊ/ foo-KOH; French pronunciation: [fuˈko]), or **Foucault's pendulum**, named after the French physicist Léon Foucault, is a simple device conceived as an experiment to demonstrate the rotation of the Earth. While it had long been known that the Earth rotated, the introduction of the Foucault pendulum in 1851 was the first simple proof of the rotation in an easy-to-see experiment. Today, Foucault pendulums are popular displays in science museums and universities.

A Foucault pendulum installed at the California Academy of Sciences. The Earth's rotation causes the trajectory of the pendulum to change over time, knocking down pins at different positions as time elapses and the Earth rotates

Index

absolute space, 23, 26
acceptor, 122
adiabatic, 140, 141, 212
Angstrom, ix
angular momentum, 18, 19, 21, 28, 29, 46, 48, 50, 53, 171, 196, 223, 227, 230, 233
antenna, 92
aperture, 102, 104, 150, 151
apogee, 198
array, 156
asymptote, x, 48
asymptotic, 3
attenuation, 95

beam, 11
Bessel, 131
"besselh", 167
"besselj", 114, 167
binary pair, 206
binding energy, 168
black hole, 218, 223, 227
body frame, 51, 54
Born approximation, 173
bosons, 170
boundary conditions, 64, 65, 71, 75, 76, 79, 80, 99, 103, 133, 135, 137, 149, 150, 193, 240
bound states, 235
Brite-Wigner, 8
burn time, 44

C++, viii
calibrate, 238

carriers, 122
Casimir effect, 192
central forces, 18, 21, 48
chaotic, 1, 30, 31, 34
charge to mass ratio, 59, 60
circular aperture, 151
collinear points, 204, 205
command line, viii, 303
compressibility, 109
Compton scattering, 10, 11
conduction, 116–120, 124, 126
    band, 116
conductor, 78
confinement, 84
conservative forces, 61
continuum, 175
contour, 64
coordinate time, 232, 234, 235
Coriolis, 23, 25
correspondence principle, 166
cross section, 20, 174
current loop, 73
cutoff frequency, 97, 99
cylinder, 68

damping, 30, 31, 145, 148, 160–162
"daughters", 191
Debye, 105
deep, 114–117, 128, 168, 186
degenerate states, 171
demo, 303
density of states, 117
depletion, 121
deuterium, 88, 89

dialogue, 12
dielectric constant, 75, 76, 78–82
"diff", 61
diffraction, 150
diffusion, 109, 121, 124, 127, 128, 130, 135
    coefficient, 128
diode, 120
dipole antenna, 92, 156
dipole radiated angular distribution, 90
dipole radiation, 90
dispersion, 95, 100, 114, 115, 117
dissipation, 96, 148, 150
distance of closest approach, 22
donor, 122
doping, 124
Doppler, 143, 144, 181, 183, 206
    broadening, 181
drag, 40
driven, 30, 103, 133, 146, 147, 158, 160, 162
    oscillator, 158
drum, 148–150
"dsolve", 3, 5, 24, 146, 160, 192, 222, 232, 241

eccentricity, 48, 49, 198, 199
"eig", 142
eigenfunctions, 142–144, 172, 174, 186–188
eigenvalues, 142
Einstein, 105
electric dipole decay, 180
electron, 59, 60, 110, 116–121, 124, 172, 175–177, 182, 186, 188, 193, 215–218
ellipse, 229
EMF, 82–85
energy band, 189
energy density, 218
energy gap, 118
entropy, 108, 109
equations of motion, 38
equatorial, 230, 232–234, 236
equilibrium solution, 137

equipotential, 38, 64, 80, 81, 204
escape velocity, 42
Euler angle, 51
Euler transformation, 11, 12
eV, ix
event horizon, 231
extrasolar planet, 206

"factor", 65
feature size, 193
Fermi level, 117
Fermi pressure, 215, 217, 218
fermions, 170
fft, 17
flow, 102
"fminsearch", 39–41, 168, 189
Foucault, 24, 308
Fourier, 71
    transform, 14
four-vector, 144
frame dragging, 231, 233, 235
freeze, 128, 129
freezing, 128
fusion, 84–86, 88, 89, 212, 215, 244

Gaussian, 8
general relativity, 36
geodesic, 222–226, 228, 231–234
Google, 307
"gradient", 61
gravity assist, 200
gravity wave, 236
GUI, 303

Hamiltonian matrix, 186
Hammer, 150
Hankle, 167
heat, 128–130, 133–138, 141, 213
    equation, 133
Heisenberg uncertainty, 163
Helmholtz, 65
Hermite, 164
"hist", 9
holes, 120
home wiring, 101
Hubble parameter, 244

ideal gas law, 210, 211
identical particles, 170
"ifft", 17
image charge, 75
impact parameter, 20–22, 223–225, 228, 229
impurity states, 120
induced EMF, 83
inertial system, 33
initial conditions, 1, 29, 30, 123, 132, 133, 135, 160, 208, 223, 232
"integral", 106
intercept, 4
intuition, x
"inv", 191
inverse, 14
ionic core, 109
ionic crystal, 109, 111, 181
ITER, 88

junction, 118
Jupiter, 200

Kepler, 48
Kerr metric, 230, 233
Kerr rotation, 232
Koenig-Penny, 188

Lagrange points, 202
Lagrangian, 34, 35
Lane-Emden, 210, 211
Laplace, 14–17, 68, 71, 72
    transform, 14
latitude, 25
launch window, 198
least action, 40, 41
lifetime, 179
light deflection, 223
limiting case, 27
line width, 179
Lissajous, 156
look back time, 246
Lorentz, 144

magnetic bottle, 84
magnetic field, 59, 65, 73–76, 83–86, 97, 98
Malthus, 3
Maple, 164
Maxwell-Boltzmann, 88, 89, 102, 108
membranes, 159
metric, 224, 226, 228, 230, 233
"min", 129
missile, 4–6
molecule, 108
moment of inertia, 53
momentum, 12
Monte Carlo, 5, 9, 11, 12, 102, 108
movies, 36, 84, 158
multiple scattering, 13, 14, 33, 51
muon, 11

natural frequency, 145
natural width, 180, 181
neutron, 218
Newton, 36, 38
nodal lines, 159, 160
non-central, 48
    force, 48
number density, 119, 120, 215, 216, 220
numerical methods, 307
nutation, 56, 58

Ode, 45, 26, 28, 35, 41, 47, 50, 57, 85, 122, 123, 139, 205, 207, 211, 217, 228
Ohms law, 99
Oppenheimer-Volkoff-Tolman, 240, 241, 243
optimization, 44

palette, 303
payload, 42–44, 198
"pdepe", 133
pendulum, 23–34
perigee, 198

perihelion, 228
periodic potential, 181, 188
perturbed solutions, 172
phonon, 105
photoelectric effect, 175
photon orbit, 226, 227
piano, 150
Planck, ix
plasma, 84, 86, 88, 90
plot3, 47
plumbing, 43, 44
Poisson equation, 120
polytropic, 209
population growth, 3
precession, 53–55, 58
pressure, 110
proper time, 227
property editor, 303
psec, ix

"quad", 89
quadrupole, 66, 68–70
quarter wave, 92, 156
"quiver", 80

radial geodesic, 226
radiation length, 33
radiation pressure, 215
rand, 7
random walk, 32
range, 7, 9, 10, 13, 30, 34, 66, 101, 110, 156, 213, 228, 237, 238
reception time, 244
redshift, 233, 244–247
re-entrant, 25
    orbit, 230, 237
relative phase, 92–94, 156
relaxation time, 135
resonances, 183
rest energy, 244
rigid body, 51, 53
Roche, 202, 203
rocket, 40
    stages, 280

scattering, 20, 173
Schrödinger, ix
    equation, 166, 185, 186
Schwarzschild radius, 223
script, x
semiconductor, 116
series, 62
shallow, 114, 115, 205
skin depth, 99
slider, 303
slinky, 139
solar constant, 218
solar data, 211
solar system, 206
"solve", 21
"sort", 202
sound velocity, 148
space frame, 51, 52, 57
species, 191
specific heat, 105–107, 140
standing wave, 160
stark effect, 171
stellar pressure, 238
stellar pulsation, 213
step driver, 145
stochastic, 32, 33
Sun, 84
supernova, 244
surface charge, 82
symbolic math, viii
symbolic solution, 3, 223

target, 4–6, 19, 177
temperature, 128
test body, 237
"thin" barriers, 183, 184
Thomson, J. J., 60
three body orbit, 207
tidal forces, 237
tide, 36
top, 53
trajectory, 1
transfer orbit, 196

transforms, 14
transient, 31
transverse electric, 97, 98
two stage, 43

valence band, 116
viscosity, 103
    of air, 41, 113

water wave, 114–117
waveguide, 96

wavelength, 220
waves, 114
white dwarf, 215
Wikipedia, 307
wobble, 206
wrapper, 303

Young's modulus, 140

World Scientific Publishing Co., Inc

27 Warren Street, Suite 401-402
Hackensack, NJ 07601, USA
Tel: (201) 487-9655 Fax: (201) 487-9656
Federal I.D. No. 13-3401101

INVOICE

ACCOUNT NO : U000000002

Sold To:

WORLDPAY USA
UNITED STATES

Ship To:

DAVID DEMSKE
12063 LITTLE PATUXENT PKWY
COLUMBIA   MD 21044
UNITED STATES

PAGE :	1
INVOICE NO :	20375245
INVOICE DATE :	19/05/15

OUR REF:	CUSTOMER REF:	TERMS	SPECIAL INSTRUCTIONS	SHIP VIA
WEB02463	WP26460	07 DAYS	THIS IS A PREPAID VISA CARD ORDER THAT HAS ALREADY BEEN CHARGED.	U26:UPS GRD RE

ISBN	TITLE	QUANTITY ORDERED	SHIPPED	BACKORDER STATUS	PUB. DATE (MM/YY)	PRICE	DIS%	AMOUNT US$
9789814623940 9330 S	MORE PHY MATLAB [W/ MEDIA PACK] PLS REFER TO INVOICE FOR REF. ONLY. VISA#: XXXX-XXXX-XXXX3744 EXP: 2/2016	1	1	0		35.00		35.00

GOODS TOTAL	35.00
FREIGHT/POSTAGE	.00
SUB-TOTAL	35.00
N.J. TAX %	.00

Payment Mode : - By Cheque to the office address at the top of the page
                - By Funds Transfer to the bank address below

Payment payable to : World Scientific Publishing Co. Inc

Bank	Address	Account No
Bank Of America	208 Harristown Road,	000251294962

NYP = Not-yet-published.   O/P = Out-of-print.
O/S = Out-of-stock.   TOS = Temporary-out-of-stock.

NOTE: Back Order/Orders will be invoiced & shipped when published. Please note pub. date.

For payment and future queries, please quote our Invoice No. Error should be notified immediately.

If you prefer to receive invoices in electronic format, please provide us with your email address.

Routing No	TOTAL VALUE
021200339	35.00

Glen Rock, NJ 07452

This is a computer generated invoice, no signature is required

Checks should be crossed and made payable to
World Scientific Publishing Co., Inc